Quadrangular Algebras

Quadrangular Algebras

Richard M. Weiss

Mathematical Notes 46

PRINCETON UNIVERSITY PRESS

PRINCETON AND OXFORD

Library of Congress Cataloging-in-Publication Data

Weiss, Richard M. (Richard Mark), 1946-
        Quadrangular algebras / Richard M. Weiss.
                p. cm. – (Mathematical Notes; 46)
        Includes index.
        ISBN-13: 978-0-691-12460-5 (pbk. : alk. paper)
        ISBN-10: 0-691-12460-4 (pbk. : alk. paper)
            1. Forms, Quadratic. 2. Algebra. I. Title. II. Mathematical notes
        (Princeton University Press) ; 46.

QA243.W45 2005
512.7'4—dc22

                                2005046572

British Library Cataloging-in-Publication Data is available

The publisher would like to acknowledge the author of this
volume for providing the camera-ready copy from which this
book was printed

Printed on acid-free paper.

pup.princeton.edu

Printed in the United States of America

10 9 8 7 6 5 4 3 2 1

Contents

Preface   vii

Chapter 1. Basic Definitions   1

Chapter 2. Quadratic Forms   11

Chapter 3. Quadrangular Algebras   21

Chapter 4. Proper Quadrangular Algebras   29

Chapter 5. Special Quadrangular Algebras   37

Chapter 6. Regular Quadrangular Algebras   45

Chapter 7. Defective Quadrangular Algebras   59

Chapter 8. Isotopes   77

Chapter 9. Improper Quadrangular Algebras   83

Chapter 10. Existence   95

Chapter 11. Moufang Quadrangles   109

Chapter 12. The Structure Group   125

Bibliography   133

Index   134

Preface

The classical groups, for which there is no canonical definition, may be taken to be the linear groups on right vector spaces of finite dimension over a field or skew field and the subgroups of linear groups on right vector spaces of arbitrary dimension leaving invariant a non-degenerate hermitian or skew-hermitian *pseudo-quadratic form* of finite Witt index. The notion of a pseudo-quadratic form was introduced by Jacques Tits in Chapter 8 of [11] to allow a characteristic-free description of these groups. To each classical group (as defined here) there is associated, via the notion of a BN-pair, a spherical building.

The classification of irreducible spherical buildings of rank at least three in [11][1] showed that the only such buildings are those arising from

    (i) a classical group,
    (ii) an isotropic K-form of an exceptional algebraic group or
    (iii) a group of "mixed type."

The existence of isotropic K-forms of exceptional algebraic groups (and the corresponding spherical buildings) is intimately connected with the existence of certain classes of non-associative algebras, in particular, *alternative division rings* and *quadratic Jordan division algebras of degree three*.

The definition of an *alternative division ring* can be obtained from the definition of a skew field by keeping certain identities which hold in a skew field and then deleting the associativity of multiplication.

The definition of a *quadratic Jordan division algebra of degree three* (called, more succinctly, an *hexagonal system* in [12]) can be obtained by taking a cubic separable field extension E/K, keeping certain identities involving the norm and trace as well as the structure of E as a vector space over K and then deleting the multiplication on E (and all the axioms which involve this multiplication).

In both cases, it is geometry (the connection to projective planes in the first case, to Moufang hexagons in the second) which guided the choice of identities to be kept.[2]

---

[1] Subsequently extended in [12] to the case of spherical buildings of rank two, also called *generalized polygons*, satisfying the Moufang condition.

[2] In fact, quadratic Jordan algebras were introduced by McCrimmon who was building on work of Jacobson and Albert going back to [7], i.e. to quantum mechanics rather than geometry. What is nevertheless true is that the notion of an hexagonal system would have turned up in Tits's work on Moufang hexagons (done at a time when he was a frequent visitor at Yale) even if these other developments had never taken place.

In this book, we introduce a new class of non-associative algebras whose definition is obtained by an analogous process. This time we start with the definition of an anisotropic pseudo-quadratic space over a skew field L with non-trivial involution σ such that either

(i) L is a commutative field or
(ii) L is a quaternion division algebra and σ is its standard involution.

We keep certain identities and the structure of L as a vector space with a quadratic form over the set K of traces (a subfield of L) and then discard the multiplication on L (and all the axioms which involve this multiplication).

The definition of these "quadrangular algebras"—the choice of which identities to keep and which structures to discard—is derived from the study of Moufang quadrangles; more precisely, from the proof of Theorem 21.12 of [12]. This is the situation which leads to the quadrangles of type E_6, E_7, E_8 and F_4.

The quadrangles of type E_6, E_7 and E_8 are, as their names suggest, the spherical buildings of certain K-forms of groups of type E_6, E_7 and E_8. More precisely, quadrangles of type E_ℓ are the spherical buildings of K-groups having index

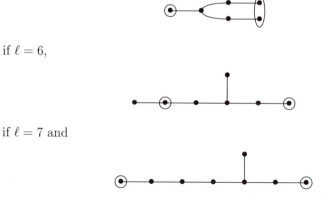

if $\ell = 6$,

if $\ell = 7$ and

if $\ell = 8$.

In Chapters 1–9 of this book we give the complete classification of quadrangular algebras. In particular, we show that a quadrangular algebra is either *special, improper, regular* (but not special), or *defective* (but not improper).

The *special* quadrangular algebras are the quadrangular algebras which arise from an anisotropic pseudo-quadratic space over a pair (L, σ) as in (i) or (ii) above.

The quadrangular algebras which are *regular* but not special exist, it turns out, only in dimensions 8, 16 and 32, and the corresponding quadratic forms are precisely those (in dimensions 6, 8 and 12) known to classify the K-forms of E_6, E_7 and E_8 indicated above. These K-forms are, in fact, classified by isotopy classes of quadrangular algebras which are regular but not special.

More precisely, a maximal unipotent subgroup of one of these forms is defined by commutator relations expressed in terms of a quadrangular algebra Ξ in the corresponding isotopy class, and the anisotropic kernel is precisely the structure group of Ξ. All of these things are explained in Chapters 10–12.

By [10] and Chapter 41 (see Figures 11–14) of [12], all the non-split forms of exceptional groups of K-rank greater than one, other than the three indicated above, are classified by either alternative division algebras[3] or hexagonal systems.[4] In this sense, quadrangular algebras belong in a series with alternative and Jordan division algebras.

The remaining quadrangular algebras exist only in characteristic two. *Improper* quadrangular algebras are of two types, one classified by degenerate anisotropic quadratic spaces and the other by indifferent sets, the structures which give rise to (and classify) the Moufang quadrangles of indifferent type.

The most exotic quadrangular algebras are those which are *defective* but not improper. These exist, it turns out, only in dimension four, not, however, over the original field K, but instead over a certain field F extending K (and which might even be of infinite dimension over K) such that F/K is purely inseparable. These quadrangular algebras classify the Moufang quadrangles of type F_4 which were discovered in the course of working out the original proof of Theorem 21.12 in [12].

Those parts of the classification of quadrangular algebras presented here in Chapters 1–7 were obtained to a large extent by extracting the purely algebraic parts of Chapters 26–28 in [12] (i.e. the proof of Theorem 21.12), supplying algebraic arguments along the way to repair the numerous gaps left by this strategy. This distillation of the purely algebraic parts of the classification of the exceptional Moufang quadrangles yields a much clearer and more satisfying understanding of these remarkable geometrical structures.

By combining the classification of quadrangular algebras with just the first few pages of Chapter 26 of [12], we obtain a new version of the proof of Theorem 21.12. This is explained in Chapter 11.

We emphasize that the classification of quadrangular algebras given in Chapters 1–9 is completely elementary and (except for a few standard facts about Clifford algebras cited in Chapter 2) self-contained.

In [3] Tom De Medts introduced a class of algebras he called "quadrangular systems" which parametrize all six families of Moufang quadrangles, not just the exceptional quadrangles considered here, and gave a reorganization of the classification of all Moufang quadrangles (Chapters 21–28 in [12]) based on this unifying concept. Our notion of a quadrangular algebra, which focuses only on the exceptional quadrangles, is quite different.

As indicated above, we hope that quadrangular algebras will turn out to be of interest alongside alternative and Jordan algebras. A first step toward a more general theory might be to investigate quadrangular algebras which do not satisfy the last axiom, the one requiring a certain "quasi-pseudo-

---

[3]More precisely, those which are quadratic in the sense of (20.1) and (20.3) in [12].

[4]Classified by Racine and Petersson (building on years of work by Jacobson, McCrimmon and others) in [8] and [9]; see Chapters 15 and 30 in [12].

quadratic form" (a frightening term which we will, in fact, avoid) to be anisotropic.

This book was written while the author was a guest of the University of Würzburg and while he was supported by a Research Prize from the Humboldt Foundation. He would like to express his gratitude to both of these institutions. It is also a pleasure to thank Yoav Segev, whose comments and suggestions led to many improvements in the text.

Chapter One

Basic Definitions

Before we can give the definition of a quadrangular algebra (in 1.17 below), we need to review a few standard notions.

Definition 1.1. A *quadratic space* is a triple (K, L, q), where K is a (commutative) field, L is a vector space over K and q is a quadratic form on L, that is, a map from L to K such that

(i) $q(u + v) = q(u) + q(v) + f(u, v)$ and
(ii) $q(tu) = t^2 q(u)$

for all $u, v \in L$ and all $t \in K$, where f is a bilinear form on L (i.e. a symmetric bilinear map from $L \times L$ to K). A quadratic space (K, L, q) is called *anisotropic* if

$$q(u) = 0 \text{ if and only if } u = 0.$$

A *basepoint* of a quadratic space (K, L, q) is an element 1 of L such that

$$q(1) = 1.$$

A *pointed quadratic space* is a quadratic space with a distinguished basepoint. Suppose that $(K, L, q, 1)$ is a pointed quadratic space (with basepoint 1). The *standard involution* of $(K, L, q, 1)$ is the map σ from L to itself given by

$$(1.2) \qquad\qquad u^\sigma = f(1, u)1 - u$$

for all $u \in L$, where f is as in (i) above. We set

$$(1.3) \qquad\qquad v^{-1} = v^\sigma / q(v)$$

for all v such that $q(v) \neq 0$. We will always identify K with its image under the map $t \mapsto t \cdot 1$ from K to L.

Note that if q, f and σ are as in 1.1 then $\sigma^2 = 1$ and $q(u^\sigma) = q(u)$ for all $u \in L$. It follows that

$$(1.4) \qquad\qquad f(u^\sigma, v^\sigma) = f(u, v)$$

for all $u, v \in L$ as well as

$$(1.5) \qquad\qquad q(v^{-1}) = q(v)^{-1}$$

and

$$(1.6) \qquad\qquad (v^{-1})^{-1} = v$$

for all $v \in L^*$.

Definition 1.7. Let L be an arbitrary ring. An *involution* of L is an anti-automorphism σ of L such that $\sigma^2 = 1$. For each involution σ of L, we set

$$L_\sigma = \{u + u^\sigma \mid u \in L\}.$$

The elements of L_σ are called *traces* (with respect to σ).

Note that in 1.2 we have called σ the standard involution of the pointed quadratic space $(K, L, q, 1)$ even though it is not, according to 1.7, really an involution (since there is no multiplication on L). The next two definitions will make it clear why we have done this.

Definition 1.8. Let E/K be a separable quadratic field extension, let N denote its norm and let σ denote the non-trivial element of $\mathrm{Gal}(E/K)$, so $N(u) = uu^\sigma$ for all $u \in E$. Let $\alpha \in K^*$ and let $M(2, E)$ denote the K-algebra of 2×2 matrices over E. The *quaternion algebra* $(E/K, \alpha)$ is the subalgebra

$$\left\{ \begin{bmatrix} a & \alpha b^\sigma \\ b & a^\sigma \end{bmatrix} \mid a, b \in E \right\}$$

of $M(2, E)$. Let $L = (E/K, \alpha)$.

We identify E (and thus also $K \subset E$) with its image in L under the map

$$u \mapsto \begin{bmatrix} u & 0 \\ 0 & u^\sigma \end{bmatrix}$$

and set

$$e = \begin{bmatrix} 0 & \alpha \\ 1 & 0 \end{bmatrix}.$$

Then every element of L can be written uniquely in the form $u + ev$ with $u, v \in E$ and multiplication in L is determined by associativity, the distributive laws and the identities $e^2 = \alpha$ and $ue = eu^\sigma$ for all $u \in E$. The extension of $\sigma \in \mathrm{Gal}(E/K)$ to the map from L to itself (which we also denote by σ) given by

(1.9) \qquad\qquad\qquad $(u + ev)^\sigma = u^\sigma - ev$

for all $u, v \in E$ is an involution of L (as defined in 1.7) called the *standard involution* of L. Both the center $Z(L)$ of L and the set of traces L_σ (as defined in 1.7) equal K and $ww^\sigma \in K$ for all $w \in L$. The extension of the norm N to a map from L to K (which we also denote by N) given by

$$N(w) = ww^\sigma$$

for all $w \in L$ is a quadratic form on L (as a vector space over K) called the *reduced norm* of L. An element w of L is invertible if and only if $N(w) \neq 0$, in which case

(1.10) \qquad\qquad\qquad $w^{-1} = w^\sigma/N(w).$

In particular, L is a division algebra if and only if the reduced norm N is anisotropic (as defined in 1.1). Since

(1.11) \qquad\qquad\qquad $N(u + ev) = N(u) - \alpha N(v)$

for all $u, v \in E$, it follows that L is a division algebra (i.e. a skew field) if and only if $\alpha \notin N(E)$.

Definition 1.12. Let L be a skew field, let σ be an involution of L and let $K = L_\sigma$ (as defined in 1.7). We will say that the pair (L, σ) is *quadratic* if either L is commutative and $\sigma \neq 1$ (in which case K is a subfield of L, L/K is a separable quadratic extension and σ is the unique non-trivial element in $\mathrm{Gal}(L/K)^1$) or L is quaternion and σ is its standard involution as defined in 1.9 (in which case $K = Z(L)$). Suppose that (L, σ) is quadratic and let

$$(1.13) \qquad\qquad q(u) = uu^\sigma$$

for all $u \in L$ (so q is the norm of the extension L/K if L is commutative and q is the reduced norm of L if L is quaternion). Then $(K, L, q, 1)$ is a pointed anisotropic quadratic space and

$$(1.14) \qquad\qquad f(u, v) = uv^\sigma + vu^\sigma$$

for all $u, v \in L$, where f is as in 1.1.i. By 1.2, it follows that σ is the standard involution of $(K, L, q, 1)$ and by 1.10, the element u^{-1} defined in 1.3 is, in fact, the inverse of u in the skew field L for all $u \in L^*$.

The following classical result is attributed (but only in characteristic different from two) in [5] (page 187) to J. Dieudonné. The general case can be found in Theorem 2.1.10 of [4]. We mention this result only to indicate that it is well known that among skew fields with involution, those singled out in 1.12 are exceptional.

Theorem 1.15. *Let L be a skew field with a non-trivial involution σ and let L_σ be as in 1.7. Then either L is generated by L_σ as a ring or the pair (L, σ) is quadratic as defined in 1.12.*

Definition 1.16. A (skew-hermitian) *pseudo-quadratic space* is a set

$$(L, \sigma, X, h, \pi),$$

where L is a skew field, σ is an involution of L (as defined in 1.7), X is a right vector space over L, h is a skew-hermitian form on X (i.e. h is a bi-additive map from $X \times X$ to L such that

(i) $h(a, bu) = h(a, b)u$ and
(ii) $h(a, b)^\sigma = -h(b, a)$

for all $a, b \in X$ and all $u \in L$) and π is a map from X to L such that

(iii) $\pi(a + b) \equiv \pi(a) + \pi(b) + h(a, b) \pmod{L_\sigma}$ and
(iv) $\pi(au) \equiv u^\sigma \pi(a) u \pmod{L_\sigma}$

for all $a, b \in X$ and all $u \in L$, where L_σ is as in 1.7. A pseudo-quadratic space

$$(L, \sigma, X, h, \pi)$$

is called *anisotropic* if

---

[1]Let $F = \{u \in L \mid u^\sigma = u\}$. Then F is a subfield of L and L/F is a separable quadratic extension. Since the map $x \mapsto x + x^\sigma$ from L to F is linear over F and not the 0-map, it is onto. Thus $K = F$.

(v) $\pi(a) \equiv 0 \pmod{L_\sigma}$ only if $a = 0$.

and *standard* if

(iv$'$) $\pi(au) = u^\sigma \pi(a)u$

for all $a \in X$ and all $u \in L$.

Let

$$\Pi = (L, \sigma, X, h, \pi)$$

be an arbitrary pseudo-quadratic space. By (11.28) and (11.31) of [12], there exists a map $\hat{\pi}$ from X to L such that

(i) $\hat{\pi}(a) \equiv \pi(a) \pmod{L_\sigma}$ for all $a \in X$ (so

$$\hat{\Pi} = (L, \sigma, X, h, \hat{\pi})$$

is also a pseudo-quadratic space which is, in an obvious sense, equivalent to Π) and

(ii) $\hat{\Pi}$ is standard (as defined in 1.16.iv$'$).

Here now is the main definition of this monograph.

Definition 1.17. A *quadrangular algebra* is a set

$$(K, L, q, 1, X, \cdot, h, \theta),$$

where $(K, L, q, 1)$ is a pointed anisotropic quadratic space as defined in 1.1, X is a non-trivial vector space over K, $(a, v) \mapsto a \cdot v$ is a map from $X \times L$ to X (which is denoted below, and, in general, simply by juxtaposition), h is a map from $X \times X$ to L and θ a map from $X \times L$ to L satisfying the following twelve axioms (in which f is as in 1.1.i, σ is as in 1.2 and v^{-1} is as in 1.3):

(A1) The map $\cdot$ is bilinear (over K).
(A2) $a \cdot 1 = a$ for all $a \in X$.
(A3) $(av)v^{-1} = a$ for all $a \in X$ and all $v \in L^*$.

(B1) h is bilinear (over K).
(B2) $h(a, bv) = h(b, av) + f(h(a, b), 1)v$ for all $a, b \in X$ and all $v \in L$.
(B3) $f(h(av, b), 1) = f(h(a, b), v)$ for all $a, b \in X$ and all $v \in L$.

(C1) For each $a \in X$, the map $v \mapsto \theta(a, v)$ is linear (over K).
(C2) $\theta(ta, v) = t^2 \theta(a, v)$ for all $t \in K$, all $a \in X$ and all $v \in L$.
(C3) There exists a function g from $X \times X$ to K such that

$$\theta(a + b, v) = \theta(a, v) + \theta(b, v) + h(a, bv) - g(a, b)v$$

for all $a, b \in X$ and all $v \in L$.
(C4) There exists a function ϕ from $X \times L$ to K such that

$$\theta(av, w) = \theta(a, w^\sigma)^\sigma q(v) - f(w, v^\sigma)\theta(a, v)^\sigma \\ + f(\theta(a, v), w^\sigma)v^\sigma + \phi(a, v)w$$

for all $a \in X$ and $v, w \in L$.

(D1) Let $\pi(a) = \theta(a, 1)$ for all $a \in X$. Then
$$a\theta(a, v) = (a\pi(a))v$$
for all $a \in X$ and all $v \in L$ and

(D2) $\pi(a) \equiv 0 \pmod{K}$ if and only if $a = 0$ (where K has been identified with its image under the map $t \mapsto t \cdot 1$ from K to L).

As was indicated in the introduction, the definition of a quadrangular algebra is derived from the definition of an anisotropic pseudo-quadratic space defined over a skew field with involution (L, σ) which is quadratic as defined in 1.12. Roughly speaking, the axioms A1–D2 capture the structure which remains when the multiplication on L is discarded. We make these comments more precise with the following result:

Proposition 1.18. *Let*
$$\Pi = (L, \sigma, X, h, \pi)$$
be a standard anisotropic pseudo-quadratic space as defined in 1.16, suppose that the pair (L, σ) is quadratic as defined in 1.12 and let $K = L_\sigma$. Let $\cdot$ denote the scalar multiplication from $X \times L$ to X (so X is a vector space over K with respect to the restriction of $\cdot$ to $X \times K$), let q be as in 1.12 (i.e. q is the norm of L/K if L is commutative and q is the reduced norm of L if L is quaternion) and let
$$\theta(a, u) = \pi(a)u$$
for all $a \in X$ and all $u \in L$. Then
$$(K, L, q, 1, X, \cdot, h, \theta)$$
is a quadrangular algebra with ϕ identically zero (where ϕ is as in axiom C4).

Proof. As observed in 1.12, $(K, L, q, 1)$ is an anisotropic pointed quadratic space and σ is its standard involution (as defined in 1.1). Axioms A1–A3 hold since X is a right vector space over L, $K \subset Z(L)$ and $v^{-1} = v^\sigma/q(v)$ in the inverse of v in the skew field L for all $v \in L^*$.

Let f denote the bilinear form associated with q. Since $q(u^\sigma) = q(u)$ for all $u \in L$, we have $q(u) = uu^\sigma = u^\sigma u$ and thus (by 1.1.i)

(1.19) $f(u, v) = u^\sigma v + v^\sigma u = uv^\sigma + vu^\sigma$

for all $u, v \in L$. By 1.16.i and 1.16.ii (and 1.7), we have

(1.20)
$$h(av, b) = -h(b, av)^\sigma = -(h(b, a)v)^\sigma$$
$$= -v^\sigma h(b, a)^\sigma = v^\sigma h(a, b)$$

for all $a, b \in X$ and all $v \in L$. Thus B1 holds by 1.16.i, 1.20 (since σ acts trivially on $K \subset L$) and the assumption in 1.16 that h is bi-additive. Moreover,

$$h(a, bv) - h(b, av) = (h(a, b) - h(b, a))v \quad \text{by 1.16.i}$$
$$= (h(a, b) + h(a, b)^\sigma)v \quad \text{by 1.16.ii}$$
$$= f(h(a, b), 1)v \quad \text{by 1.19}$$

and

$$f(h(av,b),1) = h(av,b) + h(av,b)^\sigma \qquad \text{by 1.19}$$
$$= v^\sigma h(a,b) + (v^\sigma h(a,b))^\sigma \qquad \text{by 1.20}$$
$$= v^\sigma h(a,b) + h(a,b)^\sigma v \qquad \text{by 1.7}$$
$$= f(h(a,b),v) \qquad \text{by 1.19}$$

for all $a, b \in X$ and all $v \in L$. Thus axioms B2 and B3 hold. Axioms C1 and D1 are obvious. Axiom C2 holds by 1.16.iv$'$ (and the assumption that Π is standard) since $K \subset Z(L)$ and σ acts trivially on K. Axiom C3 holds by 1.16.iii and axiom D2 holds by 1.16.v. It remains only to prove that C4 holds. Choose $a \in X$ and $v, w \in L$. We first observe that

$$f(\theta(a, w^\sigma), v) = f(\pi(a)w^\sigma, v)$$
$$= \pi(a)w^\sigma v^\sigma + vw\pi(a)^\sigma$$

and

$$f(\theta(a, v), w^\sigma) = f(\pi(a)v, w^\sigma)$$
$$= \pi(a)vw + w^\sigma v^\sigma \pi(a)^\sigma$$

by 1.7 and 1.19. Thus

$$f(\theta(a, w^\sigma), v) + f(\theta(a, v), w^\sigma)$$
$$= \pi(a)(w^\sigma v^\sigma + vw) + (w^\sigma v^\sigma + vw)\pi(a)^\sigma.$$

Since

$$w^\sigma v^\sigma + vw \in L_\sigma \subset Z(L),$$

it follows (by 1.19) that

$$(1.21) \qquad f(\theta(a, w^\sigma), v) + f(\theta(a, v), w^\sigma) = (\pi(a) + \pi(a)^\sigma)(w^\sigma v^\sigma + vw)$$
$$= f(\pi(a), 1) \cdot f(v, w^\sigma).$$

Therefore

$$\theta(av, w) = \pi(av)w$$
$$= v^\sigma \pi(a)vw \qquad \text{by 1.16.iv}'$$
$$= -v^\sigma \pi(a)w^\sigma v^\sigma + v^\sigma \pi(a)f(v^\sigma, w) \qquad \text{by 1.19}$$
$$= -v^\sigma \theta(a, w^\sigma)v^\sigma + v^\sigma \pi(a)f(v^\sigma, w)$$
$$= v^\sigma v\theta(a, w^\sigma)^\sigma - f(\theta(a, w^\sigma), v)v^\sigma$$
$$\qquad + v^\sigma \pi(a)f(v^\sigma, w) \qquad \text{by 1.19}$$
$$= \theta(a, w^\sigma)^\sigma q(v) + f(\theta(a, v), w^\sigma)v^\sigma$$
$$\qquad - f(\pi(a), 1)f(v, w^\sigma)v^\sigma$$
$$\qquad + f(v^\sigma, w)v^\sigma \pi(a) \qquad \text{by 1.13 and 1.21}$$
$$= \theta(a, w^\sigma)^\sigma q(v) + f(\theta(a, v), w^\sigma)v^\sigma$$
$$\qquad - f(v^\sigma, w)v^\sigma \pi(a)^\sigma \qquad \text{by 1.2 and 1.4}$$
$$= \theta(a, w^\sigma)^\sigma q(v) - f(w, v^\sigma)\theta(a, v)^\sigma$$
$$\qquad + f(\theta(a, v), w^\sigma)v^\sigma \qquad \text{by 1.7.}$$

Thus C4 holds with ϕ identically zero.   □

In Chapter 5 we will characterize the quadrangular algebras which arise from anisotropic pseudo-quadratic spaces as in 1.18; see 1.28 and 3.1.

Here are some further comments about the definition 1.17 of a quadrangular algebra:

(i) We will, in general, refer to the axioms in 1.17 by their names—A1, A2, etc.—without referring explicitly to 1.17.

(ii) We have put the axioms of a quadrangular algebra into four groups in the hope of making them easier to recall. Axioms A1–A3 establish properties of the "scalar multiplication" $(a, v) \mapsto av$ (those properties which remain after the multiplication on L is taken away). Axioms B1–B3 establish properties of the "skew-hermitian form" h. Axioms C1–C4 and D1–D2 establish properties of the "quasi-pseudo-quadratic form" π and its companion θ. Axioms D1 and D2 are somehow different from the others and have been listed separately. The "polarization" of axiom D1 given in 3.22 below will play an especially important role in the classification of quadrangular algebras.

(iii) Among the axioms of a quadrangular algebra, C4 is conspicuously less "natural"-looking than the others. In the proof of 1.18, however, we saw that C4 is just 1.16.iv$'$ re-written in a way that avoids the multiplication on L.

(iv) We emphasize that there is no multiplication on L in 1.17. Thus if $a \in X$ and $u, v \in L$, then auv can only mean $(au)v$, and we will almost always, in fact, omit the parentheses in such an expression. Similarly, $auvw$ can only mean $((au)v)w$, etc.

(v) We will write sometimes at (for $a \in X$ and $t \in K$) and sometimes ta to denote the scalar multiple of a by t. Similarly, we will write sometimes ut and sometimes tu to denote the scalar multiple of an element $u \in L$ by an element $t \in K$.

(vi) Since $\sigma^2 = 1$, axiom A3 can (with the help of A1) be rewritten to say that $auu^\sigma = au^\sigma u = aq(u)$ for all $a \in X$ and all $u \in L^*$.

(vii) Note that there are no assumptions about $\dim_K L$ or $\dim_K X$ in 1.17. In particular, these dimensions are allowed to be infinite.

(viii) The map g in C3 and the map called g in Chapter 13 of [12] (which we call, for the moment, g') are related by the formula $g(a, b) = g'(b, a)$ for all $a, b \in X$. This choice makes axiom C3 look a little nicer (but at the expense of making 11.10 below look a little less nice). The basepoint 1 is called ϵ and the vector spaces X and L are called X_0 and L_0 in [12]. We have tried to make all the other notation used in this monograph conform with that in [12].

We turn now to the question of how to define an isomorphism of quadrangular algebras. We begin with a notion we call equivalence:

Definition 1.22. Two quadrangular algebras

$$(K, L, q, 1, X, \cdot, h, \theta) \quad \text{and} \quad (\hat{K}, \hat{L}, \hat{q}, \hat{1}, \hat{X}, \hat{\cdot}, \hat{h}, \hat{\theta})$$

are *equivalent* if

$$(K, L, \ldots, \cdot) = (\hat{K}, \hat{L}, \ldots, \hat{\cdot})$$

and there exists an $\omega \in K^*$ such that

$$\hat{h}(a, b) = \omega h(a, b)$$

and

$$\hat{\theta}(a, u) \equiv \omega \theta(a, u) \pmod{\langle u \rangle}$$

for all $a, b \in X$ and all $u \in L$ (where $\langle u \rangle$ denotes the subspace spanned by u over K).

Proposition 1.23. *If $(K, \ldots, \theta)$ and $(\hat{K}, \ldots, \hat{\theta})$ are equivalent quadrangular algebras, then there exist an element $\omega \in K^*$ and a map $p \colon X \to K$ such that*

(i) $p(ta) = t^2 p(a)$ *and*
(ii) $\hat{\theta}(a, u) = \omega \theta(a, u) + p(a)u$

for all $a \in X$, all $u \in L$ and all $t \in K$.

Proof. By 1.22, there exists a map $r \colon X \times L \to K$ such that

$$\hat{\theta}(a, u) = \omega \theta(a, u) + r(a, u)u$$

for all $a \in X$ and all $u \in L$. Let $p(a) = r(a, 1)$ for all $a \in X$. By A1, A2 and D1,

$$\begin{aligned}
a\big(\omega \theta(a, u) + r(a, u)u\big) &= a\hat{\theta}(a, u) \\
&= a\hat{\pi}(a)u \\
&= a\big(\omega \pi(a) + p(a) \cdot 1\big)u \\
&= \omega a \pi(a)u + aup(a) \\
&= \omega a \theta(a, u) + aup(a)
\end{aligned}$$

(where $\hat{\pi}(a) = \hat{\theta}(a, 1)$ is as in D1) for all $a \in X$ and all $u \in L$. Thus $p(a)au = r(a, u)au$ for all $a \in X$ and all $u \in L$, again by A1. By A3, $au \neq 0$ and hence $p(a) = r(a, u)$ for all $a \in X^*$ and all $u \in L^*$. By C2, $\theta(0, u) = \hat{\theta}(0, u)$ for all $u \in L$. Hence $r(0, u) = p(0) = 0$ for all $u \in L^*$. Thus (ii) holds. By C2 again, it follows that (i) holds. $\square$

Proposition 1.24. *Let $\Xi = (K, L, q, 1, X, \cdot, h, \theta)$ be a quadrangular algebra, let $\omega \in K^*$ and let p be a map from X to K such that $p(ta) = t^2 p(a)$ for all $a \in X$ and all $t \in K$. Then $\hat{\Xi} = (\hat{K}, \ldots, \hat{\cdot}, \hat{h}, \hat{\theta})$ is also a quadrangular algebra, where*

$$(\hat{K}, \ldots, \hat{X}, \hat{\cdot}) = (K, \ldots, X, \cdot),$$
$$\hat{h}(a, b) = \omega h(a, b),$$
$$\hat{\theta}(a, u) = \omega \theta(a, u) + p(a)u$$

for all $a, b \in X$ *and all* $u \in L$. *Moreover,*

$$\hat{g}(a, b) = \omega g(a, b) + p(a) + p(b) - p(a + b),$$

$$\hat{\phi}(a, u) = \omega \phi(a, u) + p(av) - p(a)q(v)$$

for all $a, b \in X$ *and all* $u \in L$, *where* g *and* $\hat{g}$ *are as in C3 and* ϕ *and* $\hat{\phi}$ *is as in C4 (with respect to* Ξ *and* $\hat{\Xi}$*).*

Proof. By 1.4, $\hat{\Xi}$ satisfies C4 with $\hat{\phi}$ as indicated. It is clear that $\hat{\Xi}$ satisfies all the other axioms (with $\hat{g}$ as indicated). $\qquad\square$

Definition 1.25. An *isomorphism* from one pointed quadratic space

$$(K, L, q, 1)$$

to another $(\hat{K}, \hat{L}, \hat{q}, \hat{1})$ is a pair of maps (ψ_0, ψ_1) such that ψ_0 is an isomorphism from K to $\hat{K}$, ψ_1 is a ψ_0-linear[2] isomorphism from L to $\hat{L}$, $\hat{q}(\psi_1(u)) = \psi_0(q(u))$ for all $u \in L$ and $\psi_1(1) = \hat{1}$. An *isomorphism* from one quadrangular algebra $(K, \ldots, \theta)$ to another $(\hat{K}, \ldots, \hat{\theta})$ is a triple of maps (ψ_0, ψ_1, ψ_2) such that (ψ_0, ψ_1) is an isomorphism from $(K, L, q, 1)$ to $(\hat{K}, \hat{L}, \hat{q}, \hat{1})$ and ψ_2 is a ψ_0-linear isomorphism from X to $\hat{X}$ satisfying, for some $\hat{\omega} \in \hat{K}^*$,

(i) $\psi_2(a \cdot u) = \psi_2(a) \hat{\cdot} \psi_1(u)$,
(ii) $\psi_1(h(a, b)) = \hat{\omega} \hat{h}(\psi_2(a), \psi_2(b))$,
(iii) $\psi_1(\theta(a, u)) \equiv \hat{\omega} \hat{\theta}(\psi_2(a), \psi_1(u)) \pmod{\langle \psi_1(u) \rangle}$

for all $a, b \in X$ and all $u \in L$. If $K = \hat{K}$ and ψ_0 is the identity map, then we will say that the isomorphism is K-*linear* and we will write (ψ_1, ψ_2) in place of (ψ_0, ψ_1, ψ_2). Two pointed quadratic spaces (respectively, quadrangular algebras) are *isomorphic* if there is an isomorphism from one to the other.

Remark 1.26. The composition of isomorphisms of quadrangular algebras is again an isomorphism. Also the inverse of an isomorphism is an isomorphism. Note, too, that if $\Xi = (K, \ldots, \theta)$ and $\hat{\Xi} = (\hat{K}, \ldots, \hat{\theta})$ are equivalent quadrangular algebras as defined in 1.22, the triple consisting of three identity maps from K to itself, from L to itself and from X to itself is an isomorphism from Ξ to $\hat{\Xi}$.

Definition 1.27. Let $\Xi = (K, \ldots, \theta)$ be a quadrangular algebra and let f and σ be as in 1.17. Then Ξ is *proper* if $\sigma \neq 1$; Ξ is *regular* if f is nondegenerate (in which case Ξ, by 3.14 below, is automatically proper); and Ξ is *defective* if it is not regular.

In 9.2 below, we will also introduce the notion of an *improper* quadrangular algebra. (Improper is not quite simply the opposite of proper.)

Definition 1.28. A quadrangular algebra is *special* if it is isomorphic (as defined in 1.25) to a quadrangular algebra coming from a standard anisotropic pseudo-quadratic space as described in 1.18. A quadrangular algebra is *exceptional* if it is proper but not special.

---

[2] i.e. ψ_1 is additive and $\psi_1(tu) = \psi_0(t)\psi_1(u)$ for all $t \in K$ and all $u \in L$.

If U is a ring, a vector space or an additive group, we will denote by U^* the set of non-zero elements of U. If U is a multiplicative group, we will denote by U^* the set of non-identity elements of U.

Chapter Two

Quadratic Forms

In this chapter we assemble the facts and definitions from the theory of quadratic forms which we will need in the study of quadrangular algebras.

Definition 2.1. An *isomorphism* from one quadratic space (K, L, q) to another $(\hat{K}, \hat{L}, \hat{q})$ is a pair of maps (ψ_0, ψ_1) such that ψ_0 is an isomorphism from K to $\hat{K}$, ψ_1 is a ψ_0-linear isomorphism from L to $\hat{L}$ and $\hat{q}(\psi_1(u)) = \psi_0(q(u))$ for all $u \in L$. Two quadratic spaces are *isomorphic* if there is an isomorphism from one to the other. We will only use the symbol $\cong$ when $K = \hat{K}$ and there is an isomorphism (ψ_0, ψ_1) such that ψ_0 is the identity map on K (in which case we will say that ψ_1 is a K-*linear* isomorphism).

Definition 2.2. A *similarity* from one quadratic space (K, L, q) to another $(\hat{K}, \hat{L}, \hat{q})$ is an isomorphism (ψ_0, ψ_1) from (K, L, q) to $(\hat{K}, \hat{L}, \hat{\gamma}\hat{q})$ for some $\hat{\gamma} \in \hat{K}^*$. Two quadratic spaces are *similar* if there is a similarity from one to the other.

Definition 2.3. Let (K, L, q) be a quadratic space and let f be as in 1.1.i. Then

$$W^\perp = \{u \in L \mid f(u, W) = 0\}$$

for each subset W of L. The subspace $L^\perp$ is called the *radical* of f and f is called *degenerate* (respectively, *non-degenerate*) if the radical of f is non-trivial (respectively, trivial). The *defect* of (K, L, q) is the quotient space $L^\perp / L^\perp \cap q^{-1}(0)$ and (K, L, q) is called *defective* (respectively, *non-defective*) if its defect is non-trivial (respectively, trivial). Thus if (K, L, q) is anisotropic, the radical of f and the defect of (K, L, q) are the same thing.

Proposition 2.4. *Let (K, L, q) be a quadratic space which is both anisotropic and defective. Then* $\mathrm{char}(K) = 2$.

Proof. If $\mathrm{char}(K) \neq 2$ and (K, L, q) is anisotropic, then $f(u, u) = 2q(u)$ (by 1.1.i) and $2q(u) \neq 0$ for all $u \in L^*$. Hence $L^\perp$ is trivial. $\square$

Proposition 2.5. *The quadratic space (K, L, q) described in 1.12 is non-defective.*

Proof. Let $u \in L^*$ and let $M = \{w \in L \mid w^\sigma = -w\}$. Then $M \neq L$. There thus exists $v \in L$ such that the product $u^\sigma v$ does not lie in M. By 1.14, therefore, $f(u, v) \neq 0$, where f is as in 1.14. $\square$

Definition 2.6. Let

$$(K, L_1, q_1), (K, L_2, q_2), \ldots, (K, L_d, q_d)$$

be quadratic spaces (all over the same field K), let

$$L = L_1 \oplus L_2 \oplus \cdots \oplus L_d$$

and let

$$q_1 \perp q_2 \perp \cdots \perp q_d$$

denote the quadratic form on L which sends an element $(u_1, u_2, \ldots, u_d)$ of L to

$$q_1(u_1) + q_2(u_2) + \cdots + q_d(u_d).$$

The quadratic space

$$(K, L, q_1 \perp q_2 \perp \cdots \perp q_d)$$

is called the *orthogonal sum* of

$$(K, L_1, q_1), (K, L_2, q_2), \ldots, (K, L_d, q_d).$$

It will also be denoted by

$$(K, L_1, q_1) \perp (K, L_2, q_2) \perp \cdots \perp (K, L_d, q_d).$$

Proposition 2.7. *Let (K, L, q) be a quadratic space, let f as in 1.1.i and let U be a subspace of L. Suppose that $\dim_K U < \infty$ and that the restriction of f to U is non-degenerate. Then*

$$L = U \oplus U^\perp,$$

where $U^\perp$ is as defined in 2.3.

Proof. We apply induction with respect to $\dim_K U$. We can assume that $U \neq 0$. Since the restriction of f to U is non-degenerate, we can choose $u, v \in U$ such that $f(u, v) \neq 0$.

Suppose first that $\mathrm{char}(K) \neq 2$. Then $f(u, v) \neq -f(u, v)$ and hence it cannot be that $f(w, w) = 0$ for all $w \in \{u + v, u, v\}$. We can thus choose $w \in U$ such that $q(w) = f(w, w)/2 \neq 0$. Let $V = \langle w \rangle$. Then $V \cap V^\perp = 0$ and

$$x - f(x, w)w/2q(w) \in V^\perp$$

for all $x \in L$. Thus $L = V \oplus V^\perp$. Now suppose that $\mathrm{char}(K) = 2$ (so $f(x, x) = 0$ for all $x \in L$). Replacing u by $u/f(u, v)$, we can assume that $f(u, v) = 1$. This time we set $V = \langle u, v \rangle$. Then $V \cap V^\perp = 0$ and

$$x + f(x, u)v + f(x, v)u \in V^\perp$$

for all $x \in L$. Thus $L = V \oplus V^\perp$ also in this case. Hence

(2.8) $$L = V \oplus V^\perp$$

in every characteristic.

From now on, we suppose that $\mathrm{char}(K)$ is arbitrary. By 2.8, we have

(2.9) $$U = V \oplus (U \cap V^\perp).$$

It follows that

$$U^\perp = V^\perp \cap (U \cap V^\perp)^\perp$$

and that the restriction of f to $U \cap V^\perp$ is non-degenerate. By induction with respect to $\dim_K U$, therefore, we have

$$V^\perp = (U \cap V^\perp) \oplus U^\perp.$$

By 2.8 and 2.9, it follows that $L = U \oplus U^\perp$. $\qquad\qquad\square$

Proposition 2.10. *Let (K, L, q) be an anisotropic quadratic space such that*

$$\dim_K L = 2,$$

let f be as in 1.1.i, let $\{u, v\}$ be a basis of L, let

$$p(x) = q(u)x^2 - f(u, v)x + q(v) \in K[x]$$

and let E be the splitting field of $p(x)$ over K. Suppose, too, that

$$f(u, v) \neq 0$$

if $\mathrm{char}(K) = 2$ (equivalently, suppose that f is non-degenerate). Then $p(x)$ is irreducible over K, E/K is a separable quadratic extension and

$$(K, L, q) \cong (K, E, q(u)N),$$

where N denotes the norm of the extension E/K (and where $\cong$ means, we recall, that there exists a K-linear isomorphism).

Proof. Replacing q by $q(u)^{-1}q$, we may assume that $q(u) = 1$, i.e. that u is a basepoint of (K, L, q). Thus

$$p(x) = x^2 - f(u, v)x + q(v).$$

We have $p(t) = q(tu - v)$ for all $t \in K$. Since (K, L, q) is anisotropic, it follows that $p(x)$ is irreducible over K. Thus E/K is a separable quadratic extension. Let w be a root of $p(x)$ in E. Then $f(u, v) - w$ is the other root and the product of the two roots is $q(v)$. Hence

$$\begin{aligned}
N(t + sw) &= (t + sw)\big(t + s(f(u, v) - w)\big) \\
&= t^2 + f(u, v)st + s^2 q(v) \\
&= q(tu + sv)
\end{aligned}$$

for all $s, t \in K$. The unique K-linear map from L to E which sends u to 1 and v to w is thus an isomorphism from (K, L, q) to (K, E, N). $\qquad\square$

Proposition 2.11. *Let (K, L, q) be an anisotropic quadratic space. If*

$$\dim_K L > 2,$$

then K is infinite.

Proof. Suppose that $|K| = r < \infty$ and
$$\dim_K L \geq 2.$$
Let f be as in 1.1.i, let U be a 2-dimensional subspace of L and let u, v be a basis of U. Then $|U| = r^2$. Thus $f(u,v) \neq 0$ if $\mathrm{char}(K) = 2$, since otherwise q would be an injective homomorphism from U to K. By 2.10, therefore, $q|_U$ is similar to the norm of a quadratic extension E/K and by 2.7, $L = U \oplus U^{\perp}$. The norm of an extension of finite fields is surjective; see, for example, (34.1) of [12]. Hence for each $v \in U^{\perp}$, there exists $u \in U$ such that $q(u+v) = q(u)+q(v) = 0$. Since q is anisotropic, it follows that $U^{\perp} = 0$ and therefore $L = U$. $\qquad\square$

Definition 2.12. A *separable plane* is a quadratic space of the form
$$(K, E, N),$$
where E/K is a separable quadratic extension and N is its norm.

In the following E^d (for any natural number d) denotes the direct sum of d copies of E.

Definition 2.13. Let (K, L, q) be a quadratic space. Then:

(i) (K, L, q) is *of type E_6* if it is anisotropic and there exist constants $\alpha_1, \alpha_2, \alpha_3 \in K^*$ and a separable plane (K, E, N) such that
$$(K, L, q) \cong (K, E^3, \alpha_1 N \perp \alpha_2 N \perp \alpha_3 N).$$

(ii) (K, L, q) is *of type E_7* if it is anisotropic and there exist constants $\alpha_1, \alpha_2, \alpha_3, \alpha_4 \in K^*$ and a separable plane (K, E, N) such that
$$(K, L, q) \cong (K, E^4, \alpha_1 N \perp \alpha_2 N \perp \alpha_3 N \perp \alpha_4 N)$$
and
$$\alpha_1 \alpha_2 \alpha_3 \alpha_4 \notin N(E).$$

(iii) (K, L, q) is *of type E_8* if it is anisotropic and there exist constants $\alpha_1, \alpha_2, \ldots, \alpha_6 \in K^*$ and a separable plane (K, E, N) such that
$$(K, L, q) \cong (K, E^6, \alpha_1 N \perp \alpha_2 N \perp \cdots \perp \alpha_6 N)$$
and
$$-\alpha_1 \alpha_2 \cdots \alpha_6 \in N(E).$$

Examples of quadratic spaces of type E_6, E_7 and E_8 (along with various observations about these quadratic spaces) can be found in (12.32)–(12.38) of [12].

Notation 2.14. Let K be a field of characteristic two and let F be a subfield such that
$$K^2 \subset F \subset K.$$
Let $*$ denote the map from $K \times F$ to F given by $t * s = t^2 s$ and let q_F denote the map from F to K given by $q_F(s) = s$ for all $s \in F$. Then F is a vector space over K with respect to $*$ and (K, F, q_F) is an anisotropic quadratic space.

Definition 2.15. Let (K, L, q) be a quadratic space. Then (K, L, q) is *of type F_4* if the following hold:

(i) (K, L, q) is anisotropic;
(ii) $\text{char}(K) = 2$; and
(iii) (K, L, q) is similar to

$$(K, E \oplus E, \alpha_1 N \perp \alpha_2 N) \perp (K, F, q_F),$$

where F is a subfield of K such that

$$K^2 \subset F \subset K,$$

(K, F, q_F) is as in 2.14, (K, E, N) is a separable plane and α_1 and α_2 are elements of K^* such that

$$\alpha_1 \alpha_2 \in F.$$

Examples of quadratic spaces of type F_4 (and various observations about these quadratic spaces) can be found in (14.19)–(14.26) of [12].

Definition 2.16. Let (K, L, q) be a quadratic space and let f be as in 1.1.i. A *norm splitting map* of (K, L, q) (or of q) is a linear automorphism T of L such that for some $\alpha \in K$ such that $\alpha = 0$ if and only if $\text{char}(K) \neq 2$ and some $\beta \in K^*$, the following hold:

(i) $q(T(v)) = \beta q(v)$;
(ii) $f(v, T(v)) = \alpha q(v)$; and
(iii) $T^2(v) - \alpha T(v) + \beta v = 0$

for all $v \in L$.

Proposition 2.17. *Let (K, E, N) be a separable plane, let*

$$\alpha_1, \ldots, \alpha_d$$

be elements of K^* and let u be an element of E such that $u + u^\sigma = 0$ if and only if $\text{char}(K) \neq 2$, where σ denotes the unique non-trivial element of $\text{Gal}(E/K)$. Then left multiplication by u is a norm splitting map of the quadratic space

$$(K, E^d, \alpha_1 N \perp \cdots \perp \alpha_d N).$$

Proof. Let

$$q = \alpha_1 N \perp \cdots \perp \alpha_d N,$$

let f denote the corresponding bilinear form, let T denote the map from E^d to itself given by

$$T(v_1, \ldots, v_d) = (uv_1, \ldots, uv_d)$$

for all $v_1, \ldots, v_d \in E$, let $\alpha = u + u^\sigma$ and let $\beta = N(u)$. Since N is multiplicative, we have $q(T(v)) = \beta q(v)$ and

$$
\begin{aligned}
f(v, T(v)) &= q(v + T(v)) - q(v) - q(T(v)) \\
&= \big(N(1 + u) - N(1) - N(u)\big) q(v) \\
&= \alpha q(v).
\end{aligned}
$$

Since

$$u^2 - \alpha u + \beta = u^2 - (u + u^\sigma)u + u^\sigma u$$
$$= 0,$$

we have $T^2(v) - \alpha T(v) + \beta v = 0$ for all $v \in E^d$. By 2.16, therefore, T is a norm splitting map of the quadratic space (K, E^d, q). □

Proposition 2.18. *Let (K, L, q) be an anisotropic quadratic space, let T be a norm splitting map of (K, L, q), let f be as in 1.1.i, let α and β be as in 2.16, let $p(x) = x^2 - \alpha x + \beta$, let E be the splitting field of $p(x)$ over K, let $u \in L^*$ and let*

$$W = \langle u, T(u) \rangle.$$

Then

    (i) $\dim_K W = 2$ and f restricted to W is non-degenerate;
    (ii) $W = \langle v, T(v) \rangle$ for all $v \in W^*$;
    (iii) $p(x)$ is the minimal polynomial of T;
    (iv) E/K is a separable quadratic extension; and
    (v) $(K, W, q|_W) \cong (K, E, q(u)N)$, where N is the norm of the extension E/K.

Proof. By 2.16.i, $T(u) \neq 0$. Thus (i) holds by 2.16.ii (since $\alpha = 0$ if and only if $\text{char}(K) \neq 2$). Choose $v \in W^*$. By 2.16.iii, $\langle v, T(v) \rangle \subset W$. Therefore (ii) follows from (i). We have

$$q(u)x^2 - f(u, T(u))x + q(T(u)) = q(u)p(x)$$

by 2.16.ii and 2.16.iii. By 2.10, therefore, (iii), (iv) and (v) hold (since $\alpha \neq 0$ if $\text{char}(K) = 2$ and $p(T) = 0$). □

Definition 2.19. Let (K, L, q) be a quadratic space, let T be a norm splitting map of (K, L, q) and let f be as in 1.1.i. A subset $u, \ldots, v$ of L is *T-orthogonal* if the subspaces

$$\langle u, T(u) \rangle, \ldots, \langle v, T(v) \rangle$$

are pairwise orthogonal (with respect to f).

Proposition 2.20. *Let (K, L, q) be an anisotropic quadratic space, let T be a norm splitting map of (K, L, q), let f be as in 1.1.i, let $u, \ldots, v$ be a T-orthogonal subset of L^* and let*

$$W = \langle u, T(u), \ldots, v, T(v) \rangle.$$

Then:

    (i) The set $u, T(u), \ldots, v, T(v)$ is a basis of W and f restricted to W is non-degenerate.
    (ii) The set $u, \ldots, v, w$ is T-orthogonal for all $w \in W^\perp$.

Proof. By 2.18.i and 2.19, (i) holds. We have

$$f(T(u), T(v)) = \beta f(u, v)$$

for all $u, v \in L$ by 2.16.i and $T(W) = W$ by 2.16.iii. Therefore

$$f(T(W^\perp), W) = f(T(W^\perp), T(W)) = 0$$

and hence $T(W^\perp) \subset W^\perp$. Thus (ii) holds.   □

The following notion is due to N. Jacobson and K. McCrimmon [6].

Definition 2.21. Let $(K, L, q, 1)$ be a pointed quadratic space and let σ be its standard involution (as defined in 1.2). Let I denote the ideal of the tensor algebra $T(L)$ generated by (the vector 1) − (the scalar 1) and the elements

$$u \otimes u^\sigma - q(u) \cdot 1$$

for all $u \in L$. Let

$$C(q, 1) = T(L)/I.$$

The algebra $C(q, 1)$ is called the *Clifford algebra of q with basepoint 1*. By Theorem 5 of [6], the canonical embedding of L into $T(L)$ composed with the natural homomorphism from $T(L)$ to $C(q, 1)$ is an embedding of L into $C(q, 1)$. We will always identify L with its image under this embedding.

Proposition 2.22. *Let $(K, L, q, 1)$ be a pointed quadratic space, let σ denote its standard involution, let X be a vector space over K and suppose that $(a, v) \mapsto av$ is a bilinear map from $X \times L$ to X such that*

(i) $a1 = a$ and
(ii) $(au)u^\sigma = q(u)a$

for all $a \in X$ and all $u \in L$. Let ψ denote the map from L to $\mathrm{End}_K X$ given by $\psi(u) = \rho_u$, where $\rho_u(a) = au$ for all $a \in X$. Then there is a unique extension of ψ to a homomorphism (of K-algebras) from $C(q, 1)$ to $\mathrm{End}_K X$. Equivalently, there is a unique extension of the map $(a, v) \mapsto av$ to a map from $X \times C(q, 1)$ to X with respect to which X is a right $C(q, 1)$-module.

Proof. Since the map $(a, u) \mapsto au$ is bilinear, the map ψ is a K-linear homomorphism of vector spaces. There thus exists a unique extension of ψ to a homomorphism (of K-algebras) from $T(L)$ to $\mathrm{End}_K X$. By (i) and (ii), the ideal I defined in 2.21 lies in the kernel of this extension. It therefore induces a homomorphism from $C(q, 1)$ to $\mathrm{End}_K X$ extending ψ. Since $C(q, 1)$ is generated by L, this homomorphism is unique.   □

Proposition 2.23. *Let*

$$\Xi = (K, L, q, 1, X, \cdot, h, \theta)$$

be a quadrangular algebra. Then there is a unique right $C(q, 1)$-module structure on X which restricts to the map $(a, v) \mapsto a \cdot v$.

Proof. This holds by A1–A3 and 2.22. □

We close this chapter with three results about the algebra $C(q, 1)$ for quadratic forms of type E_6, E_7 and E_8. The third of these results will play a crucial role in the classification of regular quadrangular algebras.

We denote by $M(n, A)$ the ring of $n \times n$ matrices over a ring A.

Proposition 2.24. *Let (K, L, q) be a quadratic space of type E_ℓ for $\ell = 6$, 7 or 8 and suppose that 1 is a basepoint of q. Let (K, E, N) be a separable plane and let $\alpha_1, \ldots, \alpha_d$ (for $d = 3$, 4 and 6) be elements of K^* as in 2.13. If $\ell = 6$, then*

$$C(q, 1) \cong M(4, E).$$

If $\ell = 7$, then

$$C(q, 1) \cong M(4, D) \oplus M(4, D),$$

where D is the quaternion division algebra

$$(E/K, \alpha_1\alpha_2\alpha_3\alpha_4)$$

(as defined in 1.8). If $\ell = 8$, then

$$C(q, 1) \cong M(32, K) \oplus M(32, K).$$

Proof. By (12.51) of [12], $C(q, 1)$ is isomorphic to the even Clifford algebra $C_0(q)$ of q. The result holds, therefore, by (12.43) of [12]. □

We mention in passing that it was shown in [1] that the converse of 2.24 is also valid:

Proposition 2.25. *Let (K, L, q) be an anisotropic quadratic form and let 1 be a basepoint of q. Suppose that $C(q, 1)$ is isomorphic to*

$$M(4, E)$$

for some field E such that E/K is a separable quadratic extension,

$$M(4, D) \oplus M(4, D)$$

for some quaternion division algebra D with center K, respectively,

$$M(32, K) \oplus M(32, K).$$

Then (K, L, q) is a quadratic space of type E_6, E_7, respectively, E_8.

If (K, L, q) is a quadratic form of type E_7 or E_8 and 1 is a basepoint of q, then by (12.54) of [12], every reflection of (K, L, q) extends to an automorphism of $C(q, 1)$ which interchanges its two direct summands. The following result is thus a direct consequence of 2.24.

Proposition 2.26. *Let (K, L, q) be a quadratic space of type E_ℓ for $\ell = 6$, 7 or 8 and suppose that 1 is a basepoint of q. Then the following hold:*

(i) *A right $C(q,1)$-module is an irreducible right $C(q,1)$-module if and only if its dimension over K is n, where*

(2.27)
$$n = \begin{cases} 8 & \text{if } \ell = 6, \\ 16 & \text{if } \ell = 7, \\ 32 & \text{if } \ell = 8. \end{cases}$$

(ii) *If X and $\hat{X}$ are two irreducible right $C(q,1)$-modules, then there exist a linear isomorphism ψ from X to $\hat{X}$ and a K-linear automorphism S of (K, L, q) mapping 1 to itself such that*

$$\psi(av) = \psi(a)S(v)$$

for all $a \in X$ and all $v \in L$.

Proof. By 2.24 and (2.17) of [12], (i) holds; (ii) is proved in (12.56) of [12]. □

Chapter Three

Quadrangular Algebras

In this chapter and the next, we assemble the basic facts about quadrangular algebras which we will need to carry out their classification. The classification of proper quadrangular algebras is comprised of three theorems which we state here so that the reader knows where we are headed in Chapters 5–7. (We recall that the terms *proper, regular, defective* and *special* are defined in 1.27 and 1.28. Note, too, that by 2.5, special implies regular. Quadrangular algebras which are not proper are classified in Chapter 9.)

Theorem 3.1. *Let*

$$\Xi = (K, L, q, 1, X, \cdot, h, \theta)$$

be a proper quadrangular algebra such that

$$\dim_K L \leq 4.$$

Then Ξ is special.

Theorem 3.2. *Let*

$$\Xi = (K, L, q, 1, X, \cdot, h, \theta)$$

be a regular quadrangular algebra which is not special. Then (K, L, q) is a quadratic space of type E_6, E_7 or E_8 (as defined in 2.13) and Ξ is uniquely determined by the pointed quadratic space $(K, L, q, 1)$.

Theorem 3.3. *Let*

$$\Xi = (K, L, q, 1, X, \cdot, h, \theta)$$

be a defective proper quadrangular algebra. Then (K, L, q) is a quadratic space of type F_4 (as defined in 2.15) and Ξ is uniquely determined by the pointed quadratic space $(K, L, q, 1)$.

For the rest of this chapter, we fix a quadrangular algebra

$$\Xi = (K, L, q, 1, X, \cdot, h, \theta)$$

and let f, σ, g, ϕ and π be as in 1.17. We are assuming here that Ξ is completely arbitrary. In the next chapter, we will specialize to the case that Ξ is proper. (Our real interest is in proper quadrangular algebras. We are separating out the few little lemmas we can prove without this hypothesis for use in Chapter 9.)

Proposition 3.4. *Let $a \in X^*$ and $u, v \in L$. If $au = av$, then $u = v$.*

Proof. Suppose that $au = av$. Then $a(u - v) = au - av = 0$ and hence $u - v = 0$ by A1 and A3. $\qquad\square$

Proposition 3.5. *For each $u \in L^*$, the map $a \mapsto au$ is an automorphism of X.*

Proof. Choose $u \in L^*$. By 1.6 and A3, we have $a = au^{-1} \cdot u$ for all $a \in X$. Hence the map $a \mapsto au$ is surjective. It is linear by A1 and injective by A3. $\square$

Proposition 3.6. $h(a,b)^\sigma = -h(b,a)$ *for all $a, b \in X$.*

Proof. Let $a, b \in X$. Setting $v = 1$ in B2 and applying A2, we have

$$h(a,b) = h(b,a) + f(h(a,b),1) \cdot 1.$$

Therefore

$$h(a,b)^\sigma = f(h(a,b),1) \cdot 1 - h(a,b) = -h(b,a)$$

by 1.2. $\qquad\square$

Proposition 3.7. $f(h(a,bu),1) = f(h(a,b),u^\sigma)$ *for all $a, b \in X$ and all $u \in L$.*

Proof. Choose $a, b \in X$ and $u \in L$. Then

$$\begin{aligned}
f(h(a,b),u^\sigma) &= f(h(a,b)^\sigma, u) && \text{by 1.4} \\
&= -f(h(b,a),u) && \text{by 3.6} \\
&= -f(h(bu,a),1) && \text{by B3} \\
&= f(h(a,bu)^\sigma,1) && \text{by 3.6}
\end{aligned}$$

and

$$f(h(a,bu)^\sigma,1) = f(h(a,bu),1)$$

by another application of 1.4 (since $1^\sigma = 1$). $\qquad\square$

Proposition 3.8. $auv = -av^\sigma u^\sigma + af(u,v^\sigma)$ *for all $a \in X$ and all $u, v \in L$.*

Proof. Choose $a \in X$ and $u, v \in L$. By A1 and A3,

$$\begin{aligned}
a(u+v)(u+v)^\sigma &= aq(u+v) \\
&= a(q(u) + f(u,v) + q(v)) \\
&= auu^\sigma + af(u,v) + avv^\sigma.
\end{aligned}$$

By A1 alone, on the other hand,

$$a(u+v)(u+v)^\sigma = auu^\sigma + auv^\sigma + avu^\sigma + avv^\sigma.$$

It follows that $auv^\sigma = -avu^\sigma + af(u,v)$. Replacing v by v^σ, we thus obtain $auv = -av^\sigma u^\sigma + af(u,v^\sigma)$ (since $\sigma^2 = 1$). $\qquad\square$

Proposition 3.9. $auvu = auf(u,v^\sigma) - av^\sigma q(u)$ *for all $a \in X$ and all $u, v \in L$.*

Proof. Choose $a \in X$ and $u, v \in L$. Then

$$auvu = (auv)u = auf(u, v^\sigma) - av^\sigma u^\sigma u$$
$$= auf(u, v^\sigma) - av^\sigma q(u)$$

by A1, A3 and 3.8. □

Proposition 3.10. $auu = a\big(uf(u, 1) - q(u) \cdot 1\big)$ *for all* $a \in X$ *and all* $u \in L$.

Proof. Set $v = 1$ in 3.9 and apply A2. □

Proposition 3.11. *Let* $a \in X$, $v \in L$ *and* $t \in K$ *and suppose that*

$$\theta(a, v) = tv.$$

Then $a = 0$ *or* $v = 0$.

Proof. Suppose that $v \neq 0$. By A1 and D1, we have

$$(ta)v = a(tv) = a\theta(a, v) = a\pi(a)v.$$

By A1–A3, it follows (since $v \neq 0$) that $a\pi(a) = ta = a(t \cdot 1)$. Hence $a = 0$ or $\pi(a) = t \cdot 1$ by 3.4. By D2, therefore, $a = 0$. □

Proposition 3.12. $\theta(0, v) = 0$ *for all* $v \in L$.

Proof. Set $t = 0$ in C2. □

Proposition 3.13. $\dim_K L > 1$.

Proof. This holds by D2 since we assume (in 1.17) that the vector space X is non-trivial. □

Proposition 3.14. *If the quadrangular algebra* Ξ *is regular (which, by 2.4, is automatic if* char$(K) \neq 2$*), then* Ξ *is proper (as defined in 1.27).*

Proof. Suppose that Ξ is regular. Thus there exists $u \in L$ such that $f(1, u) \neq 0$. By 3.13, we can choose $v \in L^*$ such that $f(1, v) = 0$. By 1.2, $u^\sigma \neq u$ if char$(K) = 2$ and $v^\sigma = -v \neq v$ if char$(K) \neq 2$. Thus σ is not the identity map. □

Proposition 3.15. *Suppose that* char$(K) = 2$. *Then*

$$h(a, au) = g(a, a)u$$

for all $a \in X$ *and all* $u \in L$.

Proof. Choose $a \in X$ and $u \in L$. By 3.12, $\theta(0, u) = 0$. By C3 with $b = a$, it follows that $h(a, au) = g(a, a)u$. □

Proposition 3.16. *Suppose that* char$(K) = 2$. *Then*

$$f(\pi(a), 1) = g(a, a)$$

for all $a \in X$.

Proof. Choose $a \in X$ and $u \in L$. By 1.2,

$$w^\sigma \equiv w \pmod K$$

for all $w \in L$. Thus by C4,

(3.17) $\qquad \pi(au) \equiv \pi(a)q(u) + \theta(a, u)f(1, u) + f(\theta(a, u), 1)u \pmod K.$

Replacing u by $1 + u$ in 3.17, we obtain

(3.18) $\qquad \begin{aligned} \pi(a(1 + u)) &\equiv \pi(a)q(1 + u) + \theta(a, 1 + u)f(1, 1 + u) \\ &\quad + f(\theta(a, 1 + u), 1)u \pmod K. \end{aligned}$

By A1 and A2, $a(1 + u) = a + au$. By C1, $\theta(a, 1 + u) = \pi(a) + \theta(a, u)$. We also have $q(1) = 1$, $f(1, 1) = 0$ and $q(1 + u) = q(1) + f(1, u) + q(u)$. By 3.18, therefore,

$$\begin{aligned} \pi(a + au) &\equiv \pi(a)q(u) + \theta(a, u)f(1, u) + f(\theta(a, u), 1)u \\ &\quad + \pi(a) + f(\pi(a), 1)u \pmod K. \end{aligned}$$

By 3.17, it follows that

$$\pi(a + au) \equiv \pi(au) + \pi(a) + f(\pi(a), 1)u \pmod K.$$

By C3 and 3.15, on the other hand,

$$\begin{aligned} \pi(a + au) &\equiv \pi(a) + \pi(au) + h(a, au) \pmod K \\ &\equiv \pi(a) + \pi(au) + g(a, a)u \pmod K. \end{aligned}$$

Thus $f(\pi(a), 1)u + g(a, a)u \in K$. By 3.13, we can choose u so that it does not lie in K. Hence $f(\pi(a), 1) = g(a, a)$. $\qquad\square$

Proposition 3.19. *Let* $\mathrm{char}(K) = 2$. *Then*

   (i) $f(\theta(a, v), v) = f(\pi(a), 1)q(v)$;
   (ii) $\phi(a, u + v) + \phi(a, u) + \phi(a, v) = f(\theta(a, u), v) + g(au, av)$; *and*
   (iii) $f(\theta(a, u), v) = f(\theta(a, v), u) + f(\pi(a), 1)f(u, v)$

for all $a \in X$ *and all* $u, v \in L$.

Proof. Choose $a \in X$ and $u, v, w \in L$. By three applications of C4, we have

$$\begin{aligned} \theta(a(u + v), w) &+ \theta(au, w) + \theta(av, w) \\ &= \theta(a, w^\sigma)^\sigma f(u, v) + \theta(a, v)^\sigma f(w, u^\sigma) + \theta(a, u)^\sigma f(w, v^\sigma) \\ &\quad + f(\theta(a, u), w^\sigma)v^\sigma + f(\theta(a, v), w^\sigma)u^\sigma \\ &\quad + (\phi(a, u + v) + \phi(a, u) + \phi(a, v))w, \end{aligned}$$

whereas

$$\theta(a(u + v), w) + \theta(au, w) + \theta(av, w) = h(au, avw) + g(au, av)w$$

by A1 and C3. Setting $w = v^\sigma$ in these two equations, we conclude that

$$\begin{aligned} h(au, avv^\sigma) &+ g(au, av)v^\sigma = f(\theta(a, v), v)u^\sigma \\ &+ \big(f(\theta(a, u), v) + \phi(a, u + v) + \phi(a, u) + \phi(a, v)\big)v^\sigma. \end{aligned}$$

We have

$$
\begin{aligned}
h(au, avv^\sigma) &= h(au, a)q(v) && \text{by A3 and B1}\\
&= h(a, au)^\sigma q(v) && \text{by 3.6}\\
&= g(a, a)q(v)u^\sigma && \text{by 3.15}\\
&= f(\pi(a), 1)q(v)u^\sigma && \text{by 3.16.}
\end{aligned}
$$

Thus

$$
\big(f(\pi(a),1)q(v) + f(\theta(a,v),v)\big)u^\sigma
$$
$$
= \big(f(\theta(a,u),v) + g(au,av) + \phi(a,u+v) + \phi(a,u) + \phi(a,v)\big)v^\sigma.
$$

Since u and v are arbitrary, we conclude by 3.13 that

$$
f(\pi(a), 1)q(v) = f(\theta(a,v), v)
$$

and

$$
f(\theta(a,u), v) + g(au, av) = \phi(a, u+v) + \phi(a,u) + \phi(a,v).
$$

Thus (i) and (ii) hold. By (i), we have

$$
\begin{aligned}
f(\theta(a, u+v), u+v) &= f(\pi(a), 1)q(u+v)\\
&= f(\pi(a), 1)\big(q(u) + f(u,v) + q(v)\big)
\end{aligned}
$$

as well as

$$
f(\theta(a, w), w) = f(\pi(a), 1)q(w)
$$

for $w = u$ and v. By C1, therefore, (iii) holds. $\qquad\square$

Proposition 3.20. *Let* $\mathrm{char}(K) = 2$. *Then*

$$
\phi(a+b, u) + \phi(a, u) + \phi(b, u) = g(au, bu) + g(a,b)q(u)
$$

for all $a, b \in X$ *and all* $u \in L$.

Proof. Choose $a, b \in X$ and $u, w \in L$. By three applications of C4 and followed by three applications of C3, we have

$$
\begin{aligned}
\theta((a+b)u, w) &+ \theta(au, w) + \theta(bu, w)\\
= &\big(h(a, bw^\sigma)^\sigma + g(a,b)w\big)q(u)\\
&+ \big(h(a, bu)^\sigma + g(a,b)u^\sigma\big)f(w, u^\sigma)\\
&+ f(h(a, bu) + g(a,b)u, w^\sigma)u^\sigma\\
&+ \big(\phi(a+b, u) + \phi(a, u) + \phi(b, u)\big)w,
\end{aligned}
$$

whereas

$$
\begin{aligned}
\theta((a+b)u, w) &+ \theta(au, w) + \theta(bu, w)\\
&= \theta(au + bu, w) + \theta(au, w) + \theta(bu, w) && \text{by A1}\\
&= h(au, buw) + g(au, bu)w && \text{by C3.}
\end{aligned}
$$

Setting $w = u^\sigma$ in these two equations, we conclude that

$$h(au, buu^\sigma) + g(au, bu)u^\sigma$$
$$= h(a, bu)^\sigma q(u) + g(a, b)q(u)u^\sigma + f(h(a, bu), u)u^\sigma$$
$$+ \big(\phi(a + b, u) + \phi(a, u) + \phi(b, u)\big)u^\sigma.$$

We have

$$h(au, buu^\sigma) = h(au, b)q(u) = h(b, au)^\sigma q(u)$$

by A3, B1 and 3.6;

$$h(a, bu)^\sigma = h(b, au)^\sigma + f(h(a, b), 1)u^\sigma$$

by B2; and

$$f(h(a, bu), u) = f(h(a, buu^\sigma), 1) = f(h(a, b), 1)q(u)$$

by A3, B1 and 3.7. Thus

$$\phi(a + b, u) + \phi(a, u) + \phi(b, u) = g(au, bu) + g(a, b)q(u).$$

$\square$

Proposition 3.21. *Let* $\mathrm{char}(K) = 2$. *Then*

$$f(\pi(av), 1) = f(\pi(a), 1)q(v)$$

for all $a \in X$ *and all* $v \in L$.

Proof. By C4, we have

$$\pi(av) = \pi(a)^\sigma q(v) + f(1, v)\theta(a, v)^\sigma$$
$$+ f(\theta(a, v), 1)v^\sigma + \phi(a, v) \cdot 1$$

and hence

$$f(\pi(av), 1) = f(\pi(a), 1)q(v) + f(1, v)f(\theta(a, v), 1)$$
$$+ f(\theta(a, v), 1)f(v, 1)$$
$$= f(\pi(a), 1)q(v)$$

for all $a \in X$ and all $v \in L$. $\square$

We close this chapter with an identity which is derived from the axiom D1 by "polarization." It will play a crucial role at several places in the classification of quadrangular algebras.

Proposition 3.22. *Suppose that* $|K| > 2$. *Then*

$$b\theta(a, v) - b\pi(a)v = ah(a, b)v - ah(a, bv)$$

for all $a, b \in X$ *and all* $v \in L$.

Proof. Choose $a, b \in X$, $v \in L$ and $t \in K$. By A1, A2, B1, C2 and C3, we have

$$\theta(a + tb, v) = \theta(a, v) + t^2\theta(b, v) + th(a, bv) - g(a, tb)v$$

and

$$\pi(a + tb) = \pi(a) + t^2\pi(b) + th(a,b) - g(a,tb).$$

By A1, therefore,

$$(a+tb)\theta(a+tb,v) = a\theta(a,v) + t^2a\theta(b,v)$$
$$+ tah(a,bv) - avg(a,tb) + tb\theta(a,v)$$
$$+ t^3b\theta(b,v) + t^2bh(a,bv) - tbvg(a,tb)$$

and

$$(a+tb)\pi(a+tb)v = a\pi(a)v + t^2a\pi(b)v$$
$$+ tah(a,b)v - avg(a,tb) + tb\pi(a)v$$
$$+ t^3b\pi(b)v + t^2bh(a,b)v - tbvg(a,tb).$$

By three applications of D1, it follows that

$$(3.23) \quad \begin{aligned} t^2\big(a\theta(b,v) - a\pi(b)v + bh(a,bv) - bh(a,b)v\big) \\ = t\big(ah(a,b)v - ah(a,bv) - b\theta(a,v) + b\pi(a)v\big). \end{aligned}$$

Since $|K| > 2$ by hypothesis, we can choose $\alpha \in K$ such that $\alpha^2 \neq \alpha$. Substituting 1 and then α for t in 3.23, we conclude that

$$(\alpha - \alpha^2)\big(ah(a,b)v - ah(a,bv) - b\theta(a,v) + b\pi(a)v\big) = 0$$

and thus

$$b\theta(a,v) - b\pi(a)v = ah(a,b)v - ah(a,bv)$$

by the choice of α. $\qquad\qquad\square$

We will always apply 3.22 in conjunction with 2.11 (except in Chapter 9, where we will have $|K| = \infty$ for a different reason). In particular, the hypothesis $|K| > 2$ in 3.22 will not cause any trouble.

Chapter Four

Proper Quadrangular Algebras

We continue introducing the most fundamental properties of quadrangular algebras. In this chapter, we focus on *proper* quadrangular algebras (as defined in 1.27).

Definition 4.1. Let $\Xi = (K, \ldots, \theta)$ be a quadrangular algebra, let f and π be as in 1.17 and let $\delta \in L$. Then Ξ is δ-*standard* whenever the following hold:

   (i) $\delta = 1/2$ (in $K \subset L$) if $\operatorname{char}(K) \neq 2$;
   (ii) $f(1, \delta) = 1$ if $\operatorname{char}(K) = 2$; and
   (iii) $f(\pi(a), \delta) = 0$ for all $a \in X$ in all characteristics.

We will say that Ξ is *standard* if $\operatorname{char}(K) \neq 2$ and Ξ is δ-standard for $\delta = 1/2$ (the only choice).

Note that 4.1.i implies that $f(1, \delta) = 1$ also if $\operatorname{char}(K) \neq 2$. If $\operatorname{char}(K) = 2$, then (by 1.2) σ is non-trivial if and only if there exists $\delta \in L$ such that $f(1, \delta) = 1$. By 3.14, therefore, δ-standard implies proper in every characteristic. The converse of this statement holds in the following sense:

Proposition 4.2. *Let*

$$\Xi = (K, L, q, 1, X, \cdot, h, \theta)$$

be a proper quadrangular algebra and let f and π be as in 1.17. Let $\delta = 1/2$ if $\operatorname{char}(K) \neq 2$ and let δ be an arbitrary element of L such that $f(1, \delta) = 1$ if $\operatorname{char}(K) = 2$. Let

$$\hat{\theta}(a, v) = \theta(a, v) - f(\pi(a), \delta)v$$

for all $a \in X$ and all $v \in L$. Then

$$\hat{\Xi} = (K, L, q, 1, X, \cdot, h, \hat{\theta})$$

is a δ-standard quadrangular algebra which is equivalent to Ξ and $\hat{\theta}$ is the unique function with these properties.

Proof. Let $p(a) = -f(\pi(a), \delta)$ for all $a \in X$. By C2, $p(ta) = t^2 p(a)$ for all $a \in X$ and all $t \in K$. Thus $\hat{\Xi}$ is the quadrangular algebra obtained by applying 1.24 to Ξ and the map p with $\omega = 1$. It is δ-standard by 4.1.ii and equivalent (as defined in 1.22) to Ξ. Suppose that $\tilde{\theta}$ is another map from $X \times L$ to L such that

$$\tilde{\Xi} = (K, L, q, 1, X, \cdot, h, \tilde{\theta})$$

is a δ-standard quadrangular algebra equivalent to Ξ. Since $\hat{\Xi}$ and $\tilde{\Xi}$ are both equivalent to Ξ, we have $\tilde{\pi}(a) \equiv \hat{\pi}(a) \pmod{K}$ for all $a \in X$ (where $\tilde{\pi}$ and $\hat{\pi}$ are defined as in D2). Since $f(\hat{\pi}(a), \delta) = f(\tilde{\pi}(a), \delta) = 0$ for all $a \in X$ and $f(\delta, 1) = 1$, it follows that $\tilde{\pi} = \hat{\pi}$. By D1, it follows that

$$a\tilde{\theta}(a, v) = a\tilde{\pi}(a)v = a\hat{\pi}(a)v = a\hat{\theta}(a, v)$$

for all $a \in X$ and all $v \in L$. Hence $\tilde{\theta} = \hat{\theta}$ by 3.4 (since $\tilde{\theta}(0, v) = 0 = \hat{\theta}(0, v)$ for all $v \in L$ by 3.12). $\qquad\square$

For the rest of this chapter, we fix a quadrangular algebra

$$\Xi = (K, L, q, 1, X, \cdot, h, \theta)$$

which we assume to be proper. By 4.2, we can assume that Ξ is, in fact, δ-standard (for some $\delta \in L$) as defined in 4.1. We continue to let f, σ, g, ϕ and π be as in 1.17.

Proposition 4.3. $g(a, b) = f(h(a, b), \delta)$ for all $a, b \in X$.

Proof. Choose $a, b \in X$. By C3,

$$\pi(a + b) = \pi(a) + \pi(b) + h(a, b) - g(a, b) \cdot 1.$$

Therefore

$$g(a, b) = g(a, b)f(1, \delta) = f(h(a, b), \delta)$$

by 4.1. $\qquad\square$

Corollary 4.4. g is a bilinear form.

Proof. This holds by B1 and 4.3. $\qquad\square$

Proposition 4.5. *Suppose that* $\mathrm{char}(K) \neq 2$. *Then*

(i) $\theta(a, u) = \frac{1}{2}h(a, au)$;
(ii) $g(a, a) = 0$; *and*
(iii) $\phi(a, u) = 0$

for all $a \in X$ *and all* $u \in L$.

Proof. Choose $a \in X$ and $u \in L$. Setting $a = b$ in C3, we have

$$\theta(2a, u) = 2\theta(a, u) + h(a, au) - g(a, a)u.$$

By C2, therefore,

(4.6) $$2\theta(a, u) = h(a, au) - g(a, a)u.$$

In particular (by A2),

(4.7) $$2\pi(a) = h(a, a) - g(a, a) \cdot 1.$$

Since $1^\sigma = 1$, we have

$$\begin{aligned}
f(h(a, a), 1) &= f(h(a, a)^\sigma, 1) && \text{by 1.4} \\
&= -f(h(a, a), 1) && \text{by 3.6.}
\end{aligned}$$

Thus $f(h(a, a), 1) = 0$. Since $f(\pi(a), 1) = 0$ by 4.1, it follows from 4.7 that $g(a, a) = 0$. Therefore

$$\theta(a, u) = \tfrac{1}{2} h(a, au)$$

by 4.6. Thus (i) and (ii) hold. Finally, we observe that

$$
\begin{aligned}
h(av, avw) &= -h(av, aw^\sigma v^\sigma) + h(av, a) f(w, v^\sigma) \\
&\qquad\qquad \text{by B1 and 3.8} \\
&= -h(aw^\sigma, avv^\sigma) - f(h(av, aw^\sigma), 1) v^\sigma - h(a, av)^\sigma f(w, v^\sigma) \\
&\qquad\qquad \text{by B2 and 3.6} \\
&= -h(aw^\sigma, a) q(v) - f(h(av, a), w) v^\sigma - h(a, av)^\sigma f(w, v^\sigma) \\
&\qquad\qquad \text{by B1, A3 and 3.7} \\
&= h(a, aw^\sigma)^\sigma q(v) + f(h(a, av), w^\sigma) v^\sigma - h(a, av)^\sigma f(w, v^\sigma) \\
&\qquad\qquad \text{by 1.4 and 3.6}
\end{aligned}
$$

for all $v, w \in L$. By C4 and (i), therefore, ϕ is identically zero. Hence (iii) holds. $\qquad\qquad\qquad\qquad\qquad\qquad\qquad\qquad\qquad\qquad\qquad\qquad\square$

Remark 4.8. When $\operatorname{char}(K) \neq 2$, the map θ and the axioms C1–C4 in the definition of a quadrangular algebra are superfluous. In their place we could just take 4.5.i as the definition of θ and deduce C1–C4 from the other axioms as follows. Axioms C1 and C2 hold by A1 and B1. By A1, B1 and B2, we have

$$
\begin{aligned}
h(a + b, (a + b)u) &= h(a, au) + h(b, bu) + h(a, bu) + h(b, au) \\
&= h(a, au) + h(b, bu) + 2h(a, bu) + f(h(b, a), 1)u
\end{aligned}
$$

for all $a, b \in X$ and all $u \in L$. Thus C3 holds. For C4, we just repeat the last few lines in the proof of 4.5.

By 4.4 and 4.5.ii, the map g is alternating if $\operatorname{char}(K) \neq 2$. If $\operatorname{char}(K) = 2$, this is, in general, not true; see 3.16 above and 6.2 below.

Proposition 4.9. *Let* $\operatorname{char}(K)$ *be arbitrary. Then*

(i) $f(\theta(a, v), v) = f(\pi(a), 1) q(v);$
(ii) $\phi(a, u + v) + \phi(a, u) + \phi(a, v) = f(\theta(a, u), v) + g(au, av);$ *and*
(iii) $f(\theta(a, u), v) = -f(\theta(a, v), u) + f(\pi(a), 1) f(u, v)$

for all $a \in X$ *and all* $u, v \in L$.

Proof. Choose $a \in X$ and $u, v \in L$. By 3.19, we can assume that $\operatorname{char}(K) \neq 2$. Thus $f(\pi(a), 1) = 0$ by 4.1. Since

$$
\begin{aligned}
f(\theta(a, u), v) &= \tfrac{1}{2} f(h(a, au), v) & \text{by 4.5.i} \\
&= \tfrac{1}{2} f(h(av, a), u^\sigma) & \text{by B3 and 3.7} \\
&= \tfrac{1}{2} f(h(av, a)^\sigma, u) & \text{by 1.4} \\
&= -\tfrac{1}{2} f(h(a, av), u) & \text{by 3.6} \\
&= -f(\theta(a, v), u);
\end{aligned}
$$

it follows that (iii) holds. Setting $u = v$ in (iii), we obtain (i). By 4.5.iii, ϕ is identically zero. Moreover,

$$\begin{aligned}
f(\theta(a,u),v) &= \tfrac{1}{2}f(h(a,au),v) & \text{by 4.5.i} \\
&= \tfrac{1}{2}f(h(av,au),1) & \text{by B3} \\
&= -\tfrac{1}{2}f(h(au,av),1) & \text{by 1.4 and 3.6} \\
&= -g(au,av) & \text{by 4.3.}
\end{aligned}$$

Thus (ii) holds.   □

Proposition 4.10. *Suppose that* $\mathrm{char}(K) = 2$ *and let*

(4.11) $$v^* = f(\delta,v) \cdot 1 + f(1,v) \cdot \delta + v$$

for all $v \in L$. *Then*

(4.12) $$\phi(a,v) = f(\theta(a,v),v^*)$$

for all $a \in X$ *and all* $v \in L$. *In particular,*

(4.13) $$\phi(a,v) = 0$$

for all $a \in X$ *and all* $v \in \langle 1,\delta \rangle$ *and*

(4.14) $$\phi(ta,v) = \phi(a,tv) = t^2\phi(a,v)$$

for all $a \in X$, *all* $v \in L$ *and all* $t \in K$.

Proof. Choose $a \in X$ and $v \in L$. By C4 (and 1.4)

(4.15) $$\begin{aligned}
\pi(av) = \pi(a)^\sigma q(v) &+ \theta(a,v)^\sigma f(1,v) \\
&+ f(\theta(a,v),1)v^\sigma + \phi(a,v) \cdot 1.
\end{aligned}$$

We have

(4.16) $$\delta^\sigma = \delta + 1$$

by 4.1.ii. Hence

$$f(\pi(a)^\sigma,\delta) = f(\pi(a),\delta^\sigma) = f(\pi(a),1)$$

by 1.4 and 4.1.iii and thus

(4.17) $$f(\pi(a)^\sigma,\delta)q(v) = f(\theta(a,v),v)$$

by 4.9.i. Therefore

$$\begin{aligned}
\phi(a,v) &= f(\phi(a,v) \cdot 1,\delta) & \text{by 4.1.ii} \\
&= f(\pi(a)^\sigma,\delta)q(v) + f(1,v)f(\theta(a,v)^\sigma,\delta) \\
&\quad + f(\theta(a,v),1)f(v^\sigma,\delta) & \text{by 4.1.iii and 4.15} \\
&= f(\theta(a,v),v) + f(1,v)f(\theta(a,v),\delta^\sigma) \\
&\quad + f(\theta(a,v),1)f(v,\delta^\sigma) & \text{by 1.4 and 4.17} \\
&= f(\theta(a,v),v) + f(1,v)f(\theta(a,v),\delta) \\
&\quad + f(\theta(a,v),1)f(v,\delta) & \text{by 4.16} \\
&= f(\theta(a,v),v^*)
\end{aligned}$$

Thus 4.12 holds. It follows that 4.13 holds (since $v^* = 0$ if $v \in \langle 1,\delta \rangle$) and 4.14 holds (by C1 and C2).   □

Proposition 4.18. *Let* char(K) *be arbitrary. Then*

$$\phi(a+b,u) = \phi(a,u) + \phi(a,v) + g(au,bu) - g(a,b)q(u)$$

for all $a, b \in X$ and all $u \in L$.

Proof. Choose $a, b \in X$ and $u \in L$. By 3.20, we can assume that char(K) $\neq$ 2. Thus ϕ is identically zero by 4.5.iii and

$$
\begin{aligned}
g(au, bu) &= \tfrac{1}{2}f(h(au,bu),1) && \text{by 4.3}\\
&= \tfrac{1}{2}f(h(au,b),u^\sigma) && \text{by 3.7}\\
&= \tfrac{1}{2}f(h(auu^\sigma,b),1) && \text{by B3}\\
&= \tfrac{1}{2}f(h(a,b),1)q(u) && \text{by A3 and B1}\\
&= g(a,b)q(u) && \text{by 4.3 again.}
\end{aligned}
$$

$\square$

Proposition 4.19. $f(\theta(a,v),\pi(a)) = q(\pi(a))f(1,v)$ *for all $a \in X$ and all $v \in L$.*

Proof. Choose $a \in X$ and $v \in L$. We have

$$
\begin{aligned}
a\theta(a,\pi(a)) &= a\pi(a)\pi(a)\\
&= a\big(\pi(a)f(\pi(a),1) - q(\pi(a))\cdot 1\big)
\end{aligned}
$$

by D1 and 3.10. By 3.4 and 3.12, it follows that

$$\theta(a,\pi(a)) - \pi(a)f(\pi(a),1) + q(\pi(a))\cdot 1 = 0.$$

Thus

$$f(\theta(a,\pi(a)),v) = f(\pi(a),v)f(\pi(a),1) - q(\pi(a))f(1,v).$$

By 4.9.iii, on the other hand,

$$f(\theta(a,\pi(a)),v) = f(\pi(a),1)f(\pi(a),v) - f(\theta(a,v),\pi(a)).$$

Thus $f(\theta(a,v),\pi(a)) = q(\pi(a))f(1,v)$. $\square$

Proposition 4.20. $a\pi(a)^\sigma\theta(a,v) = q(\pi(a))av$ *for all $a \in X$ and all $v \in L$.*

Proof. Choose $a \in X$ and $v \in L$. Then

$$
\begin{aligned}
a\pi(a)^\sigma\theta(a,v) &= -a\theta(a,v)^\sigma\pi(a) + af(\theta(a,v),\pi(a))\\
&\qquad\qquad \text{by 3.8}\\
&= a\theta(a,v)\pi(a) - a\pi(a)f(\theta(a,v),1) + aq(\pi(a))f(1,v)\\
&\qquad\qquad \text{by A1, 1.2 and 4.19}\\
&= a\pi(a)v\pi(a) - a\pi(a)f(\theta(a,v),1) + aq(\pi(a))f(1,v)\\
&\qquad\qquad \text{by D1.}
\end{aligned}
$$

By 3.9, we have

$$a\pi(a)v\pi(a) = a\pi(a)f(\pi(a),v^\sigma) - av^\sigma q(\pi(a)),$$

so

$$a\pi(a)v\pi(a) + aq(\pi(a))f(1,v)$$
$$= a\pi(a)f(\pi(a), v^\sigma) + avq(\pi(a))$$
$$= a\pi(a)f(v,1)f(\pi(a),1) - a\pi(a)f(\pi(a),v) + avq(\pi(a))$$

by two applications of 1.2. By 4.9.iii,

$$a\pi(a)f(\theta(a,v),1) = a\pi(a)f(v,1)f(\pi(a),1) - a\pi(a)f(\pi(a),v).$$

Thus

$$a\pi(a)^\sigma\theta(a,v) = avq(\pi(a)).$$

$\square$

Proposition 4.21. $\theta(a, \theta(a,v)) = f(\pi(a),1)\theta(a,v) - q(\pi(a))v$ *for all* $a \in X$ *and all* $v \in L$.

Proof. Choose $a \in X$ and $v \in L$. Then

$$
\begin{aligned}
a\theta(a,\theta(a,v)) &= a\pi(a)\theta(a,v) && \text{by D1}\\
&= a\theta(a,v)f(\pi(a),1) - a\pi(a)^\sigma\theta(a,v) && \text{by 1.2}\\
&= a\theta(a,v)f(\pi(a),1) - avq(\pi(a)) && \text{by 4.20}\\
&= a\big(\theta(a,v)f(\pi(a),1) - vq(\pi(a))\big) && \text{by A1.}
\end{aligned}
$$

By 3.4 and 3.12, it follows that

$$\theta(a,\theta(a,v)) - \theta(a,v)f(\pi(a),1) + vq(\pi(a)) = 0.$$

$\square$

Proposition 4.22. $q(\theta(a,v)) = q(\pi(a))q(v)$ *for all* $a \in X$ *and all* $v \in L$.

Proof. Choose $a \in X$ and $v \in L$. If $\mathrm{char}(K) \neq 2$, then $f(\pi(a),1) = 0$ (by 4.1) and hence

$$
\begin{aligned}
q(\theta(a,v)) &= \tfrac{1}{2}f(\theta(a,v),\theta(a,v)) \\
&= -\tfrac{1}{2}f(\theta(a,\theta(a,v)),v) && \text{by 4.9.iii}\\
&= \tfrac{1}{2}q(\pi(a))f(v,v) = q(\pi(a))q(v) && \text{by 4.21.}
\end{aligned}
$$

We can thus suppose that $\mathrm{char}(K) = 2$. Then

$$
\begin{aligned}
q(\theta(a,v))f(\pi(a),1) &= f(\theta(a,\theta(a,v)),\theta(a,v)) && \text{by 4.9.i}\\
&= q(\pi(a))f(v,\theta(a,v)) && \text{by 4.21}\\
&= q(\pi(a))q(v)f(\pi(a),1) && \text{by 4.9.i}
\end{aligned}
$$

since $f(\theta(a,v),\theta(a,v)) = 0$. It thus suffices to assume that

$$f(\pi(a),1) = 0.$$

Then $\pi(a)^\sigma = \pi(a)$ by 1.2 and $f(\theta(a,v),v) = 0$ by 4.9.i. Thus

$$
\begin{aligned}
aq(\theta(a,v)) &= a\theta(a,v)\theta(a,v)^\sigma && \text{by A3}\\
&= a\pi(a)v\theta(a,v)^\sigma && \text{by D1}\\
&= a\pi(a)^\sigma v\theta(a,v)^\sigma \\
&= a\pi(a)^\sigma\theta(a,v)v^\sigma && \text{by 3.8}\\
&= q(\pi(a))avv^\sigma && \text{by 4.20}\\
&= q(\pi(a))q(v)a && \text{by A3}
\end{aligned}
$$

and hence $q(\theta(a,v)) = q(\pi(a))q(v)$ (by 3.12).

$\square$

Proposition 4.23. $\phi(av, v^{-1})q(v) = \phi(a, v)$ *for all* $a \in X$ *and all* $v \in L^*$.

Proof. By 4.5.iii, we can assume that $\text{char}(K) = 2$. By C4, we thus have

$$(4.24) \qquad \theta(av, v^\sigma) = \theta(a, v)^\sigma q(v) + f(\theta(a, v), v)v^\sigma + \phi(a, v)v^\sigma$$

for all $a \in X$ and all $v \in L$. Let u^* (for all $u \in L$) be as in 4.11. Then $(v^{-1})^* = (v^\sigma)^* q(v)^{-1} = v^* q(v)^{-1}$ for all $v \in L^*$ and $f(v^\sigma, v^*) = 0$ and $(v^*)^\sigma = v^*$ for all $v \in L$. Therefore

$$
\begin{aligned}
\phi(av, v^{-1})q(v) &= f(\theta(av, v^\sigma), v^*)q(v)^{-1} && \text{by C1 and 4.12}\\
&= f(\theta(a, v)^\sigma, v^*) && \text{by 4.24}\\
&= f(\theta(a, v), v^*) && \text{by 1.4}\\
&= \phi(a, v) && \text{by 4.12}
\end{aligned}
$$

for all $a \in X$ and all $v \in L^*$. $\qquad\qquad\qquad\qquad\qquad\qquad\qquad\square$

Chapter Five

Special Quadrangular Algebras

We continue to assume that

$$\Xi = (K, L, q, 1, X, \cdot, h, \theta)$$

is a δ-standard quadrangular algebra for some $\delta \in L$ (as defined in 4.1) and we continue to let f, σ, g, ϕ and π be as in 1.17. In this chapter, we assume as well that

(5.1) $\dim_K L \leq 4$.

Our goal is to show that under these assumptions, Ξ is special (as defined in 1.28).

Proposition 5.2. *Let $a \in X^*$. Then either $\dim_K L = 2$ and $L = \langle 1, \pi(a) \rangle$ or $\dim_K L = 4$. Suppose that $\dim_K L = 4$. Then*

(i) *$L = \langle 1, \pi(a), w, \theta(a, w) \rangle$ and $\langle w, \theta(a, w) \rangle = \langle 1, \pi(a) \rangle^\perp$ for all non-zero elements $w \in \langle 1, \pi(a) \rangle^\perp$ if either $\mathrm{char}(K) \neq 2$ or $\mathrm{char}(K) = 2$ and $f(\pi(a), 1) \neq 0$.*

(ii) *$L = \langle 1, \delta, \pi(a), \theta(a, \delta) \rangle$ and*

$$\langle \pi(a), \theta(a, \delta) \rangle = \langle 1, \delta \rangle^\perp$$

if $\mathrm{char}(K) = 2$ and $f(\pi(a), 1) = 0$.

Proof. By D2, $\langle 1, \pi(a) \rangle$ is 2-dimensional. We can thus assume that $\dim_K L > 2$.

Suppose first that f restricted to $\langle 1, \pi(a) \rangle$ is non-degenerate. (This is the case if and only if either $\mathrm{char}(K) \neq 2$ or $\mathrm{char}(K) = 2$ and $f(\pi(a), 1) \neq 0$.) By 2.7, we have

$$L = \langle 1, \pi(a) \rangle \oplus \langle 1, \pi(a) \rangle^\perp.$$

Let w be a non-zero element of $\langle 1, \pi(a) \rangle^\perp$. Then

$$f(\theta(a, w), 1) = -f(\pi(a), w) + f(\pi(a), 1) f(w, 1) = 0$$

and

$$f(\theta(a, w), \pi(a)) = q(\pi(a)) f(1, w) = 0$$

by 4.9.iii and 4.19. Thus $\theta(a, w) \in \langle 1, w \rangle^\perp$. By 3.11, the subspace

$$\langle w, \theta(a, w) \rangle$$

is 2-dimensional. By 5.1, we conclude that

$$1, \pi(a), w, \theta(a, w)$$

is a basis of L (over K) and $\langle w, \theta(a,w) \rangle = \langle 1, \pi(a) \rangle^{\perp}$.

Assume now that $\text{char}(K) = 2$ and $f(\pi(a), 1) = 0$. By 4.1, $f(1, \delta) = 1$, so
$$L = \langle 1, \delta \rangle \oplus \langle 1, \delta \rangle^{\perp}$$
by 2.7, and $\pi(a) \in \langle 1, \delta \rangle^{\perp}$. By 4.9.i and 4.9.iii, it follows that
$$f(\theta(a,\delta), \delta) = 0$$
and
$$f(\theta(a,\delta), 1) = f(\pi(a), \delta) = 0.$$
Thus $\theta(a, \delta) \in \langle 1, \delta \rangle^{\perp}$. By 4.19,
$$f(\theta(a,\delta), \pi(a)) = q(\pi(a)).$$
Since $q(\pi(a)) \neq 0$ (by D2), we conclude that
$$\langle \pi(a), \theta(a, \delta) \rangle$$
is 2-dimensional. By 5.1, therefore, $1, \delta, \pi(a), \theta(a, \delta)$ is a basis of L and $\langle \pi(a), \theta(a,\delta) \rangle = \langle 1, \delta \rangle^{\perp}$. $\qquad\square$

Proposition 5.3. *Let* $\dim_K L = 2$. *Then there exists a multiplication on* L *(which we denote either by* $\cdot$ *or by juxtaposition) extending the multiplication on* $K \subset L$ *such that*

(i) $(L, +, \cdot)$ *is a field with* $\sigma \in \text{Gal}(L/K)$ *and*
(ii) $auv = a(uv)$ *for all* $a \in X$ *and all* $u, v \in L$ *(where* auv *is to be interpreted, as usual, as* $(au)v$*).*

Proof. Choose $a \in X^*$ and let
$$p(x) = x^2 - f(\pi(a), 1)x + q(\pi(a)) \in K[x].$$
Since
$$p(t) = q(\pi(a) - t)$$
and $q(\pi(a) - t) \neq 0$ for all $t \in K$ by D2, the polynomial $p(x)$ is irreducible over K. Let E denote the splitting field of $p(x)$ over K and let γ denote a root of $p(x)$ in E. By 5.2, $L = \langle 1, \pi(a) \rangle$. Let ψ denote the unique K-linear isomorphism (of vector spaces) from L to E which sends 1 to 1 and $\pi(a)$ to γ. By 4.1, we have $f(\pi(a), 1) = 0$ if and only if $\text{char}(K) \neq 2$. Thus the extension E/K is separable and the map $\psi \sigma \psi^{-1}$ (composition from right to left) acts trivially on K and maps γ to
$$f(\pi(a), 1) - \gamma,$$
the other root of $p(x)$. Therefore $\psi \sigma \psi^{-1} \in \text{Gal}(E/K)$.

There is a unique K-bilinear multiplication $\cdot$ on L such that
$$\pi(a) \cdot \pi(a) = f(\pi(a), 1)\pi(a) - q(\pi(a)) \cdot 1$$
and $1 \cdot u = u \cdot 1 = u$ for all $u \in L$. We endow L with this multiplication. Then ψ is multiplicative, so L is a field and ψ is, in fact, an isomorphism. Thus (i) holds. By 3.10,
$$b\pi(a)\pi(a) = b(\pi(a)\pi(a))$$
for all $b \in X$. By A1, therefore,
$$buv = b(uv)$$
for all $b \in X$ and all $u, v \in L$. Thus (ii) holds.

Lemma 5.4. *Suppose that* $\dim_K L = 4$. *Let* $a \in X^*$ *and let*

$$X_a = \{b \in X \mid b\theta(a, v) = b\pi(a)v \text{ for all } v \in L^*\}.$$

Then there exists a multiplication on L *(which we denote either by* $\cdot$ *or by juxtaposition) extending the multiplication on* $K \subset L$ *such that*

(i) $(L, +, \cdot)$ *is a quaternion division algebra with center* K *and* σ *is its standard involution and*

(ii) $buv = b(uv)$ *for all* $b \in X_a$ *and all* $u, v \in L$.

Proof. We assume first that either $\text{char}(K) \neq 2$ or

$$\text{char}(K) = 2 \text{ and } f(\pi(a), 1) \neq 0$$

and then choose a non-zero element $w \in \langle 1, \pi(a) \rangle^\perp$. By 5.2.i,

$$1, \pi(a), w, \theta(a, w)$$

is a basis of L and

(5.5) $\langle w, \theta(a, w) \rangle = \langle 1, \pi(a) \rangle^\perp$

(so, in particular,

(5.6) $w^\sigma = -w$ and $\theta(a, w)^\sigma = -\theta(a, w)$).

By 4.1, we have $f(\pi(a), 1) = 0$ if and only if $\text{char}(K) \neq 2$. Let

$$p(x) = x^2 - f(1, \pi(a))x + q(\pi(a)) \cdot 1 \in K[x].$$

Since $p(t) = q(\pi(a) - t) \neq 0$ for all $t \in K$ by D2, the polynomial $p(x)$ is irreducible over K. Let E be its splitting field over K and let γ be a root of $p(x)$ in E. The extension E/K is separable. Let Q denote the quaternion algebra

$$(E/K, -q(w))$$

(as defined in 1.8). Thus $Q = \{u + ev \mid u, v \in E\}$, where

$$eu \cdot ev = -q(w)\bar{u}v,$$
$$eu \cdot v = e \cdot uv,$$
$$u \cdot ev = e \cdot \bar{u}v,$$

where $u \mapsto \bar{u}$ denotes the unique element of $\text{Gal}(E/K)$. Let ψ denote the unique K-linear isomorphism (of vector spaces) from L to Q which sends 1 to 1, $\pi(a)$ to γ, w to e and $\theta(a, w)$ to $e\bar{\gamma}$. Then $\psi\sigma\psi^{-1}$ is the standard involution of Q (i.e. the map which sends $u + ev$ to $\bar{u} - ev$ for all $u, v \in E$) and

$$N(\psi(u)) + q(w) = q(u) + q(w) = q(u + w) \neq 0$$

for all $u \in \langle 1, \pi(a) \rangle$, where N denotes the norm of the extension E/K. By 1.8, it follows that Q is a division algebra.

We now endow L with the unique K-bilinear multiplication such that

$$\pi(a)w = \theta(a, w) = w\pi(a)^\sigma,$$
$$\theta(a, w)\pi(a) = wq(\pi(a)) = \pi(a)^\sigma\theta(a, w),$$
$$\theta(a, w)w = -\pi(a)q(w),$$
$$w\theta(a, w) = -\pi(a)^\sigma q(w),$$
$$\pi(a)\pi(a)^\sigma = q(\pi(a)) \cdot 1,$$
$$w^2 = -q(w) \cdot 1,$$
$$\theta(a, w)^2 = -q(\pi(a))q(w) \cdot 1$$

and $1 \cdot u = u \cdot 1 = u$ for all $u \in L$. Then

$$L = \langle 1, \pi(a) \rangle \oplus \langle w, \theta(a, w) \rangle$$
$$= \langle 1, \pi(a) \rangle \oplus w\langle 1, \pi(a) \rangle;$$

the restriction of ψ to $\langle 1, \pi(a) \rangle$ is an isomorphism from $\langle 1, \pi(a) \rangle$ to E; and (after a bit of calculation)

$$wu \cdot wv = -q(w)u^\sigma v,$$
$$wu \cdot v = w \cdot uv \quad \text{and}$$
$$u \cdot wv = w \cdot u^\sigma v$$

for all $u, v \in \langle 1, \pi(a) \rangle$. Since $\psi(u^\sigma) = \overline{\psi(u)}$ and

$$\psi(u + wv) = \psi(u) + e\psi(v)$$

for all $u, v \in \langle 1, \pi(a) \rangle$, we conclude that ψ is an isomorphism. Thus (i) holds (in the case we are considering).

Let $b \in X_a$, so $b\theta(a, w) = b\pi(a)w$. Thus

$$
\begin{aligned}
bw\theta(a, w) &= -b\theta(a, w)^\sigma w^\sigma + bf(\theta(a, w), w^\sigma) &&\text{by 3.8} \\
&= b\theta(a, w)w^\sigma - bf(\theta(a, w), w) &&\text{by 5.6} \\
&= b\pi(a)ww^\sigma - bf(\pi(a), 1)q(w) &&\text{by 4.9.i} \\
&= b\big(\pi(a) - f(\pi(a), 1) \cdot 1\big)q(w) &&\text{by A1–A3} \\
&= -b\pi(a)^\sigma q(w) &&\text{by 1.2}
\end{aligned}
$$

and

$$b\theta(a, w)^2 = -bq(\theta(a, w)) = -bq(\pi(a))q(w)$$

by 3.10, 4.22 and 5.6. Thus $buv = b(uv)$ for $u = w$ and $v = \theta(a, w)$ and for $u = v = \theta(a, w)$. By A1–A3, 1.2, 3.8–3.10 and 5.6 (and a bit of calculation),

$$buv = b(uv)$$

holds for all other $u, v \in \{1, \pi(a), w, \theta(a, w)\}$ as well. By A1, therefore, (ii) holds.

It remains to consider the case that $\mathrm{char}(K) = 2$ and $f(\pi(a), 1) = 0$. By 5.2.ii,

$$1, \delta, \pi(a), \theta(a, \delta)$$

is a basis of L and $\langle 1, \delta \rangle^{\perp} = \langle \pi(a), \theta(a, \delta) \rangle$. In particular, σ acts trivially on $\langle \pi(a), \theta(a, \delta) \rangle)$. This time we set
$$p(x) = x^2 + x + q(\delta) \in K[x].$$
Since $p(t) = q(\delta + t) \neq 0$ for all $t \in K$, the polynomial $p(x)$ is irreducible over K. Let E denote the splitting field of $p(x)$ over K and let γ be a root of $p(x)$ in E. The extension E/K is separable. Let Q denote the quaternion algebra
$$(E/K, q(\pi(a))$$
(as defined in 1.8). Thus $Q = \{u + ev \mid u, v \in E\}$, where
$$eu \cdot ev = q(\pi(a)) \bar{u} v,$$
$$eu \cdot v = e \cdot uv,$$
$$u \cdot ev = e \cdot \bar{u} v,$$
where $u \mapsto \bar{u}$ denotes the unique element of $\mathrm{Gal}(E/K)$. Let ψ denote the unique K-linear isomorphism (of vector spaces) from L to Q which sends 1 to 1, δ to γ, $\pi(a)$ to e and $\theta(a, \delta)$ to $e\gamma$. Then $\psi \sigma \psi^{-1}$ is the standard involution of Q and
$$N(\psi(u)) + q(\pi(a)) = q(u) + q(\pi(a)) = q(u + \pi(a)) \neq 0$$
for all $u \in \langle 1, \delta \rangle$, where N denotes the norm of the extension E/K. By 1.8, it follows that Q is a division algebra.

We now endow L with the unique K-bilinear multiplication such that
$$\pi(a)\delta = \theta(a, \delta) = \delta^{\sigma} \pi(a),$$
$$\pi(a)\theta(a, \delta) = \delta q(\pi(a)),$$
$$\theta(a, \delta)\pi(a) = \delta^{\sigma} q(\pi(a)),$$
$$\theta(a, \delta)\delta^{\sigma} = \pi(a)q(\delta) = \delta\theta(a, \delta),$$
$$\delta\delta^{\sigma} = q(\delta) \cdot 1,$$
$$\pi(a)^2 = q(\pi(a)) \cdot 1,$$
$$\theta(a, \delta)^2 = q(\pi(a))q(v) \cdot 1$$
and $1 \cdot u = u \cdot 1 = u$ for all $u \in L$. Then
$$L = \langle 1, \delta \rangle \oplus \langle \pi(a), \theta(a, \delta) \rangle$$
$$= \langle 1, \delta \rangle \oplus \pi(a)\langle 1, \delta \rangle;$$
the restriction of ψ to $\langle 1, \delta \rangle$ is an isomorphism from $\langle 1, \delta \rangle$ to E; and (after a bit of calculation)
$$\pi(a)u \cdot \pi(a)v = q(\pi(a))u^{\sigma} v,$$
$$\pi(a)u \cdot v = \pi(a) \cdot uv \quad \text{and}$$
$$u \cdot \pi(a)v = \pi(a) \cdot u^{\sigma} v$$
for all $u, v \in \langle 1, \delta \rangle$. Since $\psi(u^{\sigma}) = \overline{\psi(u)}$ and $\psi(u + \pi(a)v) = \psi(u) + e\psi(v)$ for all $u, v \in \langle 1, \delta \rangle$, we conclude that ψ is an isomorphism. Thus (i) holds. By 4.19, we have
$$f(\theta(a, \delta), \pi(a)) = q(\pi(a))$$

and by 4.22, $q(\theta(a,\delta)) = q(\pi(a))q(\delta)$. By A1–A3 and 3.8–3.10 (and a bit more calculation), it follows that

$$buv = b(uv)$$

for all $b \in X_a$ and all $u, v \in \{1, \delta, \pi(a), \theta(a,\delta)\}$. By A1, therefore, (ii) holds. □

Lemma 5.7. Let $\dim_K L = 4$. Then $aLL \subset aL$ for all $a \in X$.

Proof. Choose $a \in X$ and let X_a be as in 5.4. By D1, $a \in X_a$. Thus $aLL \subset aL$ by 5.4.ii. □

Proposition 5.8. Let $\dim_K L = 4$. Then there exists a multiplication on L (which we denote either by $\cdot$ or by juxtaposition) such that

(i) $(L, +, \cdot)$ is a quaternion division algebra with center K and σ is its standard involution and
(ii) $auv = a(uv)$ for all $a \in X$ and all $u, v \in L$.

Proof. Choose $a \in X^*$, let X_a be as in 5.4 and let L be endowed with a multiplication as in 5.4. By 5.4, we just need to show that $X_a = X$.

We show first that $aL \subset X_a$. Choose $u \in L$. By D1, $a \in X_a$. Thus $\theta(a,v) = \pi(a)v$ for all $v \in L$ by 3.4. By 4.5.i if $\mathrm{char}(K) \neq 2$ and 3.15 if $\mathrm{char}(K) = 2$, we thus have

$$h(a, av) = h(a, a)v$$

for all $v \in L$. Therefore

$$\begin{aligned}
h(a, au)v &= h(a,a)u \cdot v \\
&= h(a,a) \cdot uv \\
&= h(a, a \cdot uv) = h(a, auv)
\end{aligned}$$

for all $v \in L$ since the multiplication on L is associative. By 2.11, K is infinite. By A1 and 3.22, therefore, we have

$$\begin{aligned}
au\theta(a,v) - au\pi(a)v &= ah(a, au)v - ah(a, auv) \\
&= a\big(h(a, au)v - ah(a, auv)\big) = 0,
\end{aligned}$$

so $au \in X_a$. Thus $aL \subset X_a$ as claimed.

We can assume now that $X \neq aL$. Choose $b \in X$ not contained in aL and let $v \in L$. By 3.22 again,

$$b\theta(a,v) - b\pi(a)v = ah(a,b)v - ah(a,bv).$$

By A1 and 5.7,

$$b\theta(a,v) - b\pi(a)v = bu$$

for some $u \in L$. If $u \neq 0$, then

$$\begin{aligned}
b &= \big(b\theta(a,v) - b\pi(a)v\big)u^{-1} \\
&= \big(ah(a,b)v - ah(a,bv)\big)u^{-1}
\end{aligned}$$

by A3, but

$$\big(ah(a,b)v - ah(a,bv)\big)u^{-1} \in aL$$

by 5.7. By the choice of b, we conclude that $u = 0$. Therefore

$$b\theta(a,v) - b\pi(a)v = 0,$$

i.e. $b \in X_a$.   □

We can now prove 3.1, the main result of this chapter. We restate it here:

Theorem 5.9. *Let*

$$\Xi = (K, L, q, 1, X, \cdot, h, \theta)$$

be a proper quadrangular algebra such that

$$\dim_K L \leq 4.$$

Then Ξ is special.

Proof. By 4.2, we can assume that Ξ is δ-standard (for some $\delta \in L$), so all the results of this chapter apply. By 5.2, $\dim_K L = 2$ or 4. Let L be endowed with the multiplication described in 5.3 if $\dim_K L = 2$ or in 5.8 if $\dim_K L = 4$. Thus the pair (L, σ) is quadratic as defined in 1.12 and X is a right vector space over L with respect to the map $(a, u) \mapsto au$.

By D1 and 3.4, $\theta(a, u) = \pi(a)u$ for all $a \in X$ and all $u \in L$. By C3, we have

$$\theta(a + b, u) = \theta(a, u) + \theta(b, u) + h(a, bu) - g(a, b)u$$

and

(5.10) $$\pi(a + b)u = \pi(a)u + \pi(b)u + h(a, b)u - g(a, b)u.$$

Therefore $h(a, bu) = h(a, b)u$ for all $a, b \in X$ and all $u \in L$, so 1.16.i holds. By 3.6, 1.16.ii holds. By 5.10 with $u = 1$, 1.16.iii holds.

We turn next to 1.16.iv. Choose $a \in X$ and $v \in L$. By C4, we have

$$\pi(av) \equiv \pi(a)^\sigma q(v) + \pi(a)vf(1, v) - f(\theta(a, v), 1)v \pmod{K}$$

since $w^\sigma = -w \pmod{K}$ for all $w \in L$ (by 1.2). We have

$$f(\theta(a, v), 1) = -f(\pi(a), v) + f(\pi(a), 1)f(1, v)$$

by 4.9.iii and therefore

$$\pi(a)vf(1, v) - f(\theta(a,v), 1)v = \big((\pi(a) - f(\pi(a), 1) \cdot 1)f(1, v)\big)v$$
$$+ f(\pi(a), v)v$$
$$= -\pi(a)^\sigma vf(1, v) + f(\pi(a), v)v$$

by 1.2. Since

$$\pi(a)^\sigma q(v) - \pi(a)^\sigma vf(1, v) = \pi(a)^\sigma v^\sigma v - \pi(a)^\sigma vf(1, v) \quad \text{by A3}$$
$$= \pi(a)^\sigma \big(v^\sigma - f(1, v)\big)v$$
$$= -\pi(a)^\sigma vv \qquad\qquad \text{by 1.2,}$$

we conclude that

$$\pi(av) \equiv -\pi(a)^\sigma vv + vf(\pi(a), v) \pmod{K}$$
$$\equiv v^\sigma \pi(a)v \pmod{K} \qquad \text{by 1.19.}$$

Hence 1.16.iv holds. By D2, finally, also 1.16.v holds. Thus

$$(L, \sigma, X, h, \pi)$$

is an anisotropic pseudo-quadratic space and, by 1.28, Ξ is special. $\qquad\square$

Chapter Six

Regular Quadrangular Algebras

We continue to assume that

$$\Xi = (K, L, q, 1, X, \cdot, h, \theta)$$

is a δ-standard quadrangular algebra (for some $\delta \in L$) and we continue to let f, σ, g, ϕ and π be as in 1.17. In this chapter, we assume as well that Ξ is regular (as defined in 1.27) but not special (as defined in 1.28). Our goal is to show that (K, L, q) must be a quadratic space of type E_6, E_7 or E_8 (as defined in 2.13) and that Ξ is uniquely determined by the pointed quadratic space $(K, L, q, 1)$.

Since Ξ is not special, we have

$$(6.1) \qquad\qquad \dim_K L > 4$$

by 5.9.

Proposition 6.2. *If* $\mathrm{char}(K) = 2$, *then there exists an element* $a \in X$ *such that* $f(\pi(a), 1) \neq 0$.

Proof. Suppose that $\mathrm{char}(K) = 2$ but that $f(\pi(a), 1) = 0$ for all $a \in X$. By 3.16, $g(a, a) = 0$ for all $a \in X$. By 4.4, g is bilinear. It follows that g is symmetric. On other hand,

$$
\begin{aligned}
g(b, a) &= f(h(b, a), \delta) && \text{by 4.3} \\
&= f(h(b, a)^\sigma, \delta^\sigma) && \text{by 1.4} \\
&= f(h(a, b), \delta + 1) && \text{by 4.1.ii and 3.6} \\
&= g(a, b) + f(h(a, b), 1) && \text{by 4.3}
\end{aligned}
$$

for all $a, b \in X$. Thus $f(h(a, b), 1) = 0$ for all $a, b \in X$. By B3, therefore,

$$f(h(a, b), v) = f(h(av, b), 1) = 0$$

for all $a, b \in X$ and all $v \in L$. Since Ξ is regular (by hypothesis) as defined in 1.27, it follows that h is identically zero. By 4.3, this implies that g is also identically zero.

Choose $a \in X^*$ and $u, v \in L$. By A1 and C3, we now have

$$
\begin{aligned}
\pi(a(u + v)) &= \pi(au + av) \\
&= \pi(au) + \pi(av).
\end{aligned}
$$

Applying C4 to each term, we obtain

$$(6.3) \qquad
\begin{aligned}
&\pi(a)f(u,v) + \theta(a, u)f(1, v) + \theta(a, v)f(1, u) \\
&\qquad \equiv f(\theta(a, u), 1)v + f(\theta(a, v), 1)u \pmod{K}
\end{aligned}
$$

(since $w^\sigma = w \pmod{K}$ for all $w \in L$).

Since f restricted to $\langle 1, \delta \rangle$ is non-degenerate, we have (by 2.7)

$$L = \langle 1, \delta \rangle \oplus \langle 1, \delta \rangle^\perp.$$

Thus the restriction of f to $\langle 1, \delta \rangle^\perp$ is also non-degenerate (since f itself is non-degenerate). It follows (by 6.1) that we can choose $u_1, v_1 \in \langle 1, \delta \rangle^\perp$ such that $f(u_1, v_1) = 1$.

Substituting u_1 and v_1 for u and v in 6.3, we conclude that

$$\pi(a) \in \langle 1, u_1, v_1 \rangle.$$

By D2, therefore,

$$\pi(a) \notin \langle u_1, v_1 \rangle^\perp.$$

Since f restricted to $\langle 1, \delta, u_1, v_1 \rangle$ is non-degenerate, we have

$$L = \langle 1, \delta, u_1, v_1 \rangle \oplus \langle 1, \delta, u_1, v_1 \rangle^\perp.$$

Therefore the restriction of f to

$$\langle 1, \delta, u_1, v_1 \rangle^\perp$$

is non-degenerate. It follows (again by 6.1) that we can choose

$$u_2, v_2 \in \langle 1, \delta, u_1, v_1 \rangle^\perp$$

such that $f(u_2, v_2) \neq 0$. If we now substitute u_2 and v_2 for u and v in 6.3, however, we deduce that

$$\pi(a) \in \langle 1, u_2, v_2 \rangle \subset \langle u_1, v_1 \rangle^\perp.$$

With this contradiction, we conclude that there does exist $a \in X$ such that $f(\pi(a), 1) \neq 0$. $\square$

Notation 6.4. By 6.2, we can choose $e \in X^*$ such that $f(\pi(e), 1) \neq 0$ if $\mathrm{char}(K) = 2$. If $\mathrm{char}(K) \neq 2$, let e be an arbitrary element of X^*. Let

$$u^\# = \theta(e, u)$$

for all $u \in L$ (in all characteristics).

Proposition 6.5. $u \mapsto u^\#$ is a norm splitting map of the quadratic space (K, L, q) (as defined in 2.16) whose minimal polynomial is

$$p(x) = x^2 - f(1, \pi(e))x + q(\pi(e)).$$

Proof. By 4.22, 2.16.i holds with $\beta = q(\pi(e))$. By 4.9.i, 2.16.ii holds with $\alpha = f(\pi(e), 1)$. By 4.21, 2.16.iii holds. By 4.1 and 6.4, $f(\pi(e), 1) = 0$ if and only if $\mathrm{char}(K) \neq 2$. Thus $u \mapsto u^\#$ is a norm splitting map of (K, L, q). By 2.18.iii, $p(x)$ is its minimal polynomial. $\square$

Definition 6.6. A subset $u, \ldots, v$ of L is called *e-orthogonal* if it is T_e-orthogonal as defined in 2.19, where T_e denotes the map $u \mapsto u^\#$. Thus the subset $u, \ldots, v$ is e-orthogonal if and only if the subspaces

$$\langle u, u^\# \rangle, \ldots, \langle v, v^\# \rangle$$

are pairwise orthogonal (with respect to f).

Proposition 6.7. *Let $u, \ldots, v$ be an e-orthogonal subset of L^* and let*

$$W = \langle u, u^{\#}, \ldots, v, v^{\#} \rangle.$$

Then:

   (i) The set $u, u^{\#}, \ldots, v, v^{\#}$ is a basis of W and f restricted to W is non-degenerate.

   (ii) The set $u, \ldots, v, w$ is e-orthogonal for all $w \in W^{\perp}$.

Proof. This holds by 2.20 and 6.5. $\qquad\qquad\qquad\qquad\qquad\qquad\square$

The next result is the first indication that we are in an "exceptional" situation; when Ξ is special, $eC(q, 1)$ is just the 1-dimensional subspace of X spanned by e over L.

Proposition 6.8. *Let $C(q, 1)$ be as in 2.21 and let X be endowed with the unique right $C(q, 1)$-module structure described in 2.23. Then*

$$X = eC(q, 1).$$

Proof. Let $M = eC(q, 1)$ and choose $a \in X$. By 6.1, $\dim_K L > 4$. By 6.7, therefore, we can choose $u, v \in L$ such that $1, u, v$ is an e-orthogonal set. By 2.11, we can apply 3.22. By 3.22 with e in place of a and a in place of b, we have

$$av^{\#} - a\pi(e)v \in M.$$

Thus

(6.9) $$\qquad\qquad av^{\#}u - a\pi(e)vu \in Mu = M.$$

By 3.22 with e in place of a and au in place of b, we have

(6.10) $$\qquad\qquad auv^{\#} - au\pi(e)v \in M.$$

Since $1, u, v$ is an e-orthogonal set (and $\pi(e) = 1^{\#}$), we have $av^{\#}u = -auv^{\#}$ and

$$a\pi(e)vu = -a\pi(e)uv = -au\pi(e)^{\sigma}v$$

by three applications of 3.8. Adding 6.9 and 6.10, we thus obtain

(6.11) $$\qquad\qquad au\big(\pi(e)^{\sigma} - \pi(e)\big)v \in M$$

(by A1). We have

$$\pi(e) - \pi(e)^{\sigma} = \begin{cases} 2\pi(e) & \text{if char}(K) \neq 2, \\ f(\pi(e), 1) & \text{if char}(K) = 2 \end{cases}$$

by 1.2 and 4.1. Thus, in particular, $\pi(e) - \pi(e)^{\sigma} \neq 0$ (by D2 and 6.4). By A3 and 6.11, therefore, $a \in M$. Since a is arbitrary, it follows that $M = X$. $\square$

Proposition 6.12. $h(e, eu) \neq 0$ *for all $u \in L^*$.*

Proof. Choose $u \in L^*$. If char$(K) \neq 2$, then $h(e, eu) = 2u^\#$ by 4.5.i and $u^\# \neq 0$ (by 3.11, for instance). If char$(K) = 2$, then

$$h(e, eu) = f(\pi(e), 1)u$$

by 3.15 and 3.16 and $f(\pi(e), 1) \neq 0$ by 6.4. □

Proposition 6.13. *Suppose that $1, u, v$ is an e-orthogonal subset of L (as defined in 6.6). Then*

$$h(e, euv) = 0.$$

Proof. By 3.22,

$$euv^\# - eu\pi(e)v = eh(e, eu)v - eh(e, euv).$$

On the other hand,

$$\begin{aligned}
euv^\# - eu\pi(e)v &= -ev^\#u - e\pi(e)^\sigma uv && \text{by 3.8} \\
&= -ev^\#u + e\pi(e)^\sigma vu && \text{by 3.8} \\
&= -ev^\#u - e\pi(e)vu + evuf(\pi(e), 1) && \text{by 1.2} \\
&= -2e\pi(e)vu + evuf(\pi(e), 1) && \text{by D1.}
\end{aligned}$$

Hence

(6.14) $-2e\pi(e)vu + evuf(\pi(e), 1) = eh(e, eu)v - eh(e, euv).$

If char$(K) \neq 2$, then $f(\pi(e), 1) = 0$ by 4.1 and

$$\begin{aligned}
eh(e, eu)v &= 2eu^\#v && \text{by 4.5.i} \\
&= 2e\pi(e)uv && \text{by D1} \\
&= -2e\pi(e)vu && \text{by 3.8}
\end{aligned}$$

and thus, by 6.14, $eh(e, euv) = 0$. If char$(K) = 2$, then

$$\begin{aligned}
eh(e, eu)v &= euvf(\pi(e), 1) && \text{by 3.15 and 3.16} \\
&= evuf(\pi(e), 1) && \text{by 3.8,}
\end{aligned}$$

so again $eh(e, euv) = 0$ (again by 6.14). By A3, we conclude that $h(e, euv) = 0$ (in all characteristics). □

Proposition 6.15. *Suppose that $1, u, v, w$ is an e-orthogonal subset of L. Then*

$$h(e, euvw) = 0.$$

Proof. By 3.22 and 6.13,

$$\begin{aligned}
euvw^\# - euv\pi(e)w &= eh(e, euv)w - eh(e, euvw) \\
&= -eh(e, euvw).
\end{aligned}$$

On the other hand,

$$euvw^\# - euv\pi(e)w = ew^\#uv - e\pi(e)wuv$$

by several applications of 3.8 and

$$ew^\#uv - e\pi(e)wuv = 0$$

by D1. Therefore $eh(e, euvw) = 0$. Hence $h(e, euvw) = 0$ by A3. □

Proposition 6.16. *Suppose that* $1, u, v, w, x$ *is an e-orthogonal subset of* L^*. *Then there exists* $y \in L^*$ *such that*

$$euvwx = ey$$

and the subset $1, u, v, w, x, y$ *of* L^* *is e-orthogonal.*

Proof. Let

$$\tilde{u} = \begin{cases} -u^\#/q(\pi(e)) & \text{if } \mathrm{char}(K) \neq 2, \\ u & \text{if } \mathrm{char}(K) = 2. \end{cases}$$

By 2.18.ii, $\langle \tilde{u}, \tilde{u}^\# \rangle = \langle u, u^\# \rangle$. The set

$$1, \tilde{u}, v, w, x$$

is thus also e-orthogonal. By 3.22 and 6.15,

$$evwx\tilde{u}^\# - evwx\pi(e)\tilde{u} = eh(e, evwx)\tilde{u} - eh(e, evwx\tilde{u})$$
$$= -eh(e, evwx\tilde{u}).$$

We have

$$evwx\tilde{u}^\# = -e\tilde{u}^\# vwx \qquad \text{by 3.8}$$
$$= -e\pi(e)\tilde{u}vwx \quad \text{by D1}$$

and

$$evwx\pi(e)\tilde{u} = -e\pi(e)^\sigma \tilde{u}vwx \qquad\qquad \text{by 3.8}$$
$$= e\pi(e)\tilde{u}vwx - e\tilde{u}vwxf(\pi(e), 1) \quad \text{by A1 and 1.2.}$$

Thus

(6.17) $$eh(e, evwx\tilde{u}) = 2e\pi(e)\tilde{u}vwx - e\tilde{u}vwxf(\pi(e), 1).$$

Now suppose that $\mathrm{char}(K) \neq 2$. Then $f(\pi(e), 1) = 0$ by 4.1 and $\tilde{u}^\# = u$ by C1 and 4.21. Thus

$$e\pi(e)\tilde{u} = e\tilde{u}^\# = eu$$

by D1. By 6.17, therefore,

$$euvwx = e\pi(e)\tilde{u}vwx = ey$$

for

(6.18) $$y = \tfrac{1}{2}h(e, evwx\tilde{u}).$$

Suppose that $\mathrm{char}(K) = 2$. Then $\tilde{u} = u$ and, by 6.4, $f(\pi(e), 1) \neq 0$. By 6.17, therefore,

$$euvwx = ey$$

for

(6.19) $$y = h(e, evwxu)/f(\pi(e), 1).$$

By A3, $ey = euvwx \neq 0$ and therefore $y \neq 0$ (in all characteristics).

It remains only to show that the subset

$$1, u, v, w, x, y$$

of L^* is e-orthogonal, where y is as in 6.18 or 6.19 (according to the characteristic). By 6.7.i, the restriction of f to

$$\langle 1, \pi(e), u, u^{\#}, \ldots, x, x^{\#} \rangle$$

is non-degenerate. Thus (by 2.7) there exist elements

$$\tilde{y} \in \langle 1, \pi(e), u, u^{\#}, \ldots, x, x^{\#} \rangle^{\perp}$$

and

$$y_1 \in \langle 1, \pi(e) \rangle, \ y_u \in \langle u, u^{\#} \rangle, \ldots, y_x \in \langle x, x^{\#} \rangle$$

such that

$$y = \tilde{y} + y_1 + y_u + \cdots + y_x.$$

By 6.7.ii, the subset

$$1, u, v, w, x, \tilde{y}$$

of L is e-orthogonal. It will suffice, therefore, to show that $y = \tilde{y}$. By 6.13, we have

$$h(e, e\tilde{y}u) = h(e, e\tilde{y}v) = \cdots = h(e, e\tilde{y}x) = 0.$$

By A3 and 3.8, we have

$$eyu = euvwxu$$
$$= -euuvwx = evwxq(u).$$

Thus $h(e, eyu) = 0$ by B1 and 6.15. By a similar argument,

$$h(e, eyv) = \cdots = h(e, eyx) = 0.$$

By 2.18.ii, the subset $1, u, y_v, y_w, y_x$ of L is e-orthogonal. Thus by 6.13,

$$h(e, ey_v u) = h(e, ey_w u) = h(e, ey_x u) = 0.$$

It follows (by A1 and B1) that

(6.20) $$h(e, e(y_1 + y_u)u) = h\big(e, e(y - \tilde{y} - y_v - y_w - y_x)u\big) = 0.$$

By D1 and 3.10, $e\pi(e)u$, euu and $eu^{\#}u$ all lie in eL. By A1, therefore,

$$e(y_1 + y_u)u \in e\langle 1, \pi(e), u, u^{\#} \rangle u \in eL.$$

By A3, 6.12 and 6.20, it follows that $y_1 + y_u = 0$. Therefore $y_1 = y_u = 0$ since

$$\langle 1, \pi(e) \rangle \cap \langle u, u^{\#} \rangle = 0.$$

We could equally well have argued in the last few lines with v, w or x in place of u. Thus $y_v = y_w = y_x = 0$. Hence $y = \tilde{y}$. $\square$

Proposition 6.21. $\dim_K L = 6, 8$ *or* 12.

Proof. By 2.7, 6.1, 6.7 and 6.16, we can assume that $\dim_K L > 12$ and that there exists an e-orthogonal subset

$$1, u, v, w, x, y, z$$

of L^* such that

$$euvwx = ey.$$

By 3.22 and 6.13,

(6.22)
$$\begin{aligned} eyz^{\#} - ey\pi(e)z &= eh(e, ey)z - eh(e, eyz) \\ &= eh(e, ey)z. \end{aligned}$$

We also have

$$\begin{aligned} eyz^{\#} &= euvwxz^{\#} \\ &= ez^{\#}uvwx && \text{by 3.8} \\ &= e\pi(e)zuvwx && \text{by D1} \\ &= euvwx\pi(e)z && \text{by 3.8} \\ &= ey\pi(e)z. \end{aligned}$$

By 6.22, therefore, $eh(e, ey)z = 0$. Hence by A3, $h(e, ey) = 0$. By 6.12, this contradicts the choice of y. $\qquad\square$

Proposition 6.23. *Suppose that* $\dim_K L = 12$. *Then there exists an e-orthogonal subset*

$$1, u, v, w, x, y$$

of L^* such that $euvwxy = e$.

Proof. By 6.16, there exists an e-orthogonal subset

$$1, u, v, w, x, y$$

of L^* such that $euvwx = ey$. By 2.18.ii, the subset

$$1, u, v, w, x, y^{-1}$$

is also e-orthogonal and by A3, $euvwxy^{-1} = e$. $\qquad\square$

Notation 6.24. By 2.7, 6.7, 6.21 and 6.23, we can extend the subset of L^* consisting of just 1 to an e-orthogonal subset

$$v_1, \ldots, v_d$$

of L^* with $v_1 = 1$ and $d = 3$, 4 or 6 such that

$$v_1, v_1^{\#}, \ldots, v_d, v_d^{\#}$$

is a basis of L and

(6.25)
$$ev_2v_3v_4v_5v_6 = e$$

if $d = 6$. Let $\alpha_i = q(v_i)$ for all $i \in [1, d]$ (so $\alpha_1 = 1$). Let E denote the splitting field over K of the minimal polynomial $p(x)$ (given in 6.5) of the norm splitting map $u \mapsto u^{\#}$. By 2.18.iv,v, E/K is a separable quadratic extension and

(6.26)
$$(K, L, q) \cong (K, E^d, \alpha_1 N \perp \cdots \perp \alpha_d N),$$

where N denotes the norm of the extension E/K.

Proposition 6.27. *Suppose that $d = 4$. Then $\alpha_2\alpha_3\alpha_4 \notin N(E)$, where d, $\alpha_2, \alpha_3, \alpha_4$, E and N are as in 6.24.*

Proof. Let $t = \alpha_2\alpha_3\alpha_4$ and suppose that $t = N(u)$ for some $u \in E^*$. By 2.18.v,

$$(K, \langle v_4, v_4^{\#}\rangle, q|_{\langle v_4, v_4^{\#}\rangle}) \cong (K, E, \alpha_4 N).$$

Since $\alpha_4 N(u^{-1}) = \alpha_4/t$, there thus exists an element v_4' in $\langle v_4, v_4^{\#}\rangle$ such that $q(v_4') = \alpha_4/t$. By 2.18.ii, the set v_1, v_2, v_3, v_4' is e-orthogonal. Replacing v_4 by v_4', we can thus assume that $t = 1$.

We have $ev_2v_2 = -eq(v_2) = -\alpha_2 e$ by A3,

$$f(h(ev_2, e), 1) = f(h(e, e), v_2)$$

by B3 and $h(e, e) \in \langle v_1, v_1^{\#}\rangle \subset \langle v_2\rangle^{\perp}$ by 3.15 and 4.5.i. By B1 and B2, therefore,

$$\begin{aligned} h(ev_2, ev_2) &= h(e, ev_2v_2) + f(h(ev_2, e), 1)v_2 \\ &= -\alpha_2 h(e, e). \end{aligned}$$

By similar calculations,

$$h(ev_2v_3, ev_2v_3) = -\alpha_3 h(ev_2, ev_2)$$

and

$$h(ev_2v_3v_4, ev_2v_3v_4) = -\alpha_4 h(ev_2v_3, ev_2v_3).$$

Thus

$$\begin{aligned} h(ev_2v_3v_4, ev_2v_3v_4) &= -\alpha_2\alpha_3\alpha_4 h(e, e) \\ &= -h(e, e) \end{aligned}$$

(since $t = 1$). By 3.6 and 6.15,

$$h(ev_2v_3v_4, e) = h(e, ev_2v_3v_4) = 0.$$

By B1, therefore, $h(a, a) = 0$ and $h(e, a) = h(e, e)$ for

$$a = e + ev_2v_3v_4.$$

If $\text{char}(K) \neq 2$, then

$$h(e, a) = h(e, e) = 2\pi(e) \neq 0$$

by D2 and 4.5.i, so $a \neq 0$ and thus $\pi(a) \neq 0$, again by D2; but at the same time $2\pi(a) = h(a, a) = 0$ by 4.5.i. We thus have $\text{char}(K) = 2$. By 3.15 and 3.16, therefore,

$$h(e, a) = h(e, e) = f(\pi(e), 1)$$

and thus $h(e, a) \neq 0$ by 6.4. Thus $a \neq 0$. By 4.3 and 6.15, we have $g(e, ev_2v_3v_4) = 0$. Thus

$$\pi(a) = \pi(e) + \pi(ev_2v_3v_4)$$

by C3. By three applications of C4, we have

(6.28)
$$\pi(ev_2) \equiv \pi(e)\alpha_2 \pmod{K},$$
$$\theta(ev_2, v_3) \equiv \theta(e, v_3)\alpha_2 \pmod{K} \text{ and}$$
$$\theta(ev_2, v_4) \equiv \theta(e, v_4)\alpha_2 \pmod{K}.$$

Thus
$$\theta(ev_2 v_3, v_4) \equiv \theta(ev_2, v_4)\alpha_3 \pmod{K}$$
$$\equiv \theta(e, v_4)\alpha_2\alpha_3 \pmod{K}$$

and

(6.29)
$$\pi(ev_2 v_3) \equiv \pi(ev_2)\alpha_3 \pmod{K}$$

by two more applications of C4. By one last application of C4, it follows that

$$\pi(ev_2 v_3 v_4) \equiv \pi(ev_2 v_3)\alpha_4 \pmod{K}.$$

By 6.28 and 6.29, it follows that

$$\pi(ev_2 v_3 v_4) \equiv \pi(e)\alpha_2\alpha_3\alpha_4. \pmod{K}$$
$$\equiv \pi(e) \pmod{K}$$

(since $t = 1$). Thus $\pi(a) = \pi(e) + \pi(ev_2 v_3 v_4) \in K$. By D2, however, this implies that $a = 0$. With this contradiction, we conclude that t is not contained in $N(E)$. □

Proposition 6.30. *Suppose that $d = 6$. Then $\alpha_2\alpha_3\alpha_4\alpha_5\alpha_6 = -1$.*

Proof. By 6.25,

$$ev_2 v_3 v_4 v_5 v_6 = e.$$

Thus

$$ev_2 v_3 v_4 v_5 = ev_6^{-1} = -ev_6/\alpha_6$$

by A3 and 1.3. By 3.8,

$$ev_6 v_5 v_4 v_3 v_2 = ev_2 v_3 v_4 v_5 v_6 = e.$$

Again by A3 and 1.3, it follows that

$$ev_6 = ev_2^{-1}v_3^{-1}v_4^{-1}v_5^{-1} = ev_2 v_3 v_4 v_5/\alpha_2\alpha_3\alpha_4\alpha_5.$$

Therefore

$$\alpha_2\alpha_3\alpha_4\alpha_5 ev_6 = -ev_6/\alpha_6$$

and hence $\alpha_2\alpha_3\alpha_4\alpha_5\alpha_6 = -1$. □

Proposition 6.31. *The quadratic space (K, L, q) is of type E_6, E_7 or E_8 (as defined in 2.13).*

Proof. This holds by 6.21, 6.26, 6.27 and 6.30. □

Notation 6.32. Let d and $v_1, v_2, \ldots, v_d \in L^*$ be as in 6.24 and let I denote the set of all subsets of the interval $[2, d]$, each written in ascending order. For each $x \in I$, let

$$e_x = \begin{cases} e & \text{if } x = \emptyset, \\ ev_i \ldots v_j & \text{if } x = (i, \ldots, j) \end{cases}$$

and

$$e_x^\circ = \begin{cases} e\pi(e) & \text{if } x = \emptyset, \\ e\pi(e)v_i \ldots v_j & \text{if } x = (i, \ldots, j). \end{cases}$$

Let I_m (for all $m < d$) denote the set of all subsets in I of size at most m and let

$$B = \{e_x \mid x \in I_p\} \cup \{e_x^\circ \mid x \in I_p\},$$

where $p = d - 1$ if $d = 3$ or 4 and $p = 2$ if $d = 6$. Thus

(6.33) $$|B| = \begin{cases} 8 & \text{if } d = 3, \\ 16 & \text{if } d = 4, \\ 32 & \text{if } d = 6. \end{cases}$$

Proposition 6.34. *Let*

(6.35) $$X_x = \langle e_x, e_x^\circ \rangle$$

for all $x \in I_p$, where p, I_p, e_x and e_x° are as in 6.32, and let $u \in \langle v_i, v_i^\# \rangle$ for some $i \in [1, d]$. Then

(6.36) $$X_x u = \begin{cases} X_{x \cup \{i\}} & \text{if } i \notin x, \\ X_{x \setminus \{i\}} & \text{if } i \in x \end{cases}$$

if $d = 3$ or 4;

(6.37) $$X_x u = \begin{cases} X_{x \cup \{i\}} & \text{if } i \notin x \text{ and } |x| \leq 1, \\ X_{[2,6] \setminus \{x \cup \{i\}\}} & \text{if } i \notin x \text{ and } |x| = 2, \\ X_{x \setminus \{i\}} & \text{if } i \in x \end{cases}$$

if $d = 6$; and the set B is a basis of X (over K) in all three cases.

Proof. Let

$$Y = \bigoplus_{x \in I_p} X_x.$$

Products of the form $e_x u$ and $e_x^\circ u$ can be calculated using A1–A3, D1, 3.8 and 6.25 For example, with $d = 6$,

$$\begin{aligned}
e_{(2,3)}^\circ \cdot v_4 &= ev_2 v_3 v_4 v_5 v_6 \cdot \pi(e) v_2 v_3 v_4 & \text{by 6.25} \\
&= -e\pi(e)^\sigma v_2 v_2 \cdot v_3 v_3 \cdot v_4 v_4 \cdot v_5 v_6 & \text{by 3.8} \\
&= e\pi(e)^\sigma v_5 v_6 \alpha_2 \alpha_3 \alpha_4 & \text{by A1 and A3} \\
&= \left(e_{(5,6)} f(1, \pi(e)) - e_{(5,6)}^\circ\right) \alpha_2 \alpha_3 \alpha_4 & \text{by 1.2}
\end{aligned}$$

and

$$e^{\circ}_{(2,3)} \cdot v_4^{\#} = ev_2 v_3 v_4 v_5 v_6 \cdot \pi(e) v_2 v_3 v_4^{\#} \qquad \text{by 6.25}$$

$$= ev_4^{\#} \pi(e) v_2 v_2 \cdot v_3 v_3 \cdot v_4 v_5 v_6 \qquad \text{by 3.8}$$

$$= e\pi(e) v_4 \pi(e) v_4 v_5 v_6 \alpha_2 \alpha_3 \qquad \text{by A1, A3 and D1}$$

$$= e\pi(e) \pi(e)^{\sigma} v_4 v_4 \cdot v_5 v_6 \alpha_2 \alpha_3 \qquad \text{by 3.8}$$

$$= -e_{(5,6)} q(\pi(e)) \alpha_2 \alpha_3 \alpha_4 \qquad \text{by A1 and A3.}$$

It follows from these and similar calculations that $X_x u$ is as in 6.36 and 6.37. In particular, $YL \subset Y$ by A1. By 6.8, it follows that $Y = X$. Hence B spans X. By 2.23, 2.26.i and 6.33, $\dim_K X \geq |B|$. Thus B is, in fact, a basis of X. $\square$

Proposition 6.38. *Let*

$$\hat{\Xi} = (\hat{K}, \hat{L}, \hat{q}, \hat{1}, \hat{X}, \hat{\cdot}, \hat{h}, \hat{\theta})$$

be a quadrangular algebra such that

$$(\hat{K}, \hat{L}, \hat{q}, \hat{1}, \hat{X}, \hat{\cdot}) = (K, L, q, 1, X, \cdot).$$

Then $\hat{\Xi}$ is equivalent to Ξ (as defined in 1.22).

Proof. By 4.2, we can assume that $\hat{\Xi}$, like Ξ, is δ-standard (for some $\delta \in L$). Let X_x for all $x \in I_p$ be as in 6.35, where p, I_p, e_x and e°_x are as in 6.32. By 6.34,

$$X = \bigoplus_{x \in I_p} X_x$$

and for each non-empty $x \in I_p$, there exists (at least one) $i \in [2, d]$ such that $X_x v_i = X_y$ for some $y \in I_p$ of size greater than one. Since

$$e\hat{\pi}(e) v_i \in eL = \bigoplus_{|x| \leq 1} X_x$$

for all $i \in [1, d]$ by D1 (where $\hat{\pi}(a) = \hat{\theta}(a, 1)$ for all $a \in X$), it follows that

$$e\hat{\pi}(e) \in X_{\emptyset} = \langle e, e\pi(e) \rangle.$$

By A1, D2 and 3.4, there thus exist $\omega \in K^*$ and $\beta \in K$ such that

$$\hat{\pi}(e) = \omega \pi(e) + \beta \cdot 1.$$

By 4.1.iii applied to both Ξ and $\hat{\Xi}$,

$$\beta = f(\hat{\pi}(e), \delta) - \omega f(\pi(e), \delta) = 0.$$

By A1 and D1, it follows that

$$e\hat{\theta}(e, u) = e\hat{\pi}(e) u = e(\omega \pi(e)) u = e(\omega \theta(e, u))$$

and thus

(6.39) $$\hat{\theta}(e, u) = \omega \theta(e, u)$$

for all $u \in L$ by 3.4.

By 6.6 and 6.39, the set $v_1, \ldots, v_6$ is also e-orthogonal with respect to $\hat{\theta}$. By 6.13 and 6.15, it follows that

$$\hat{h}(e, e_x) = h(e, e_x) = 0$$

for all $x \in I_p$ of cardinality is greater than one. Similarly,

$$\hat{h}(e, e_x^\circ) = h(e, e_x^\circ) = 0$$

for all $x \in I_p$ of cardinality is greater than one. (Let, for example, $x = (2, 3)$. Then $e_x^\circ = e\pi(e)v_2v_3 = ev_2^\# v_3$ by D1 and the set $1, v_2^\#, v_3$ is e-orthogonal by 2.18.ii, so $\hat{h}(e, e_x^\circ) = h(e, e_x^\circ) = 0$ by 6.13.) If $\mathrm{char}(K) \neq 2$, then

$$\hat{h}(e, eu) = \omega h(e, eu)$$

for all $u \in L$ by 4.5.i and 6.39 and if $\mathrm{char}(K) = 2$, then

$$\hat{h}(e, eu) = f(\hat{\pi}(e), 1)u$$
$$= \omega f(\pi(e), 1)u = \omega h(e, eu)$$

for all $u \in L$ by 3.15, 3.16 and 6.39. By D1, therefore $\hat{h}(e, e_x) = \omega h(e, e_x)$ and $\hat{h}(e, e_x^\circ) = \omega h(e, e_x^\circ)$ for all $x \in I_p$ of cardinality at most one. By B1 and 6.34, we conclude that

(6.40) $$\hat{h}(e, b) = \omega h(e, b)$$

for all $b \in X$.

Let

$$\tilde{h}(a, b) = \hat{h}(a, b) - \omega h(a, b)$$

for all $a, b \in X$ and let

$$V = \{a \in X \mid \hat{h}(a, b) = \omega h(a, b) \text{ for all } b \in X\}.$$

Then $e \in V$ by 6.40 and V is closed under addition by B1. Let $a \in V$ and $u \in L$. Since $a \in V$, we have $\tilde{h}(a, bu) = \tilde{h}(a, b) = 0$ for all $b \in X$. Hence

$$\tilde{h}(au, b) = -\tilde{h}(b, au)^\sigma \qquad \text{by 3.6}$$
$$= -\tilde{h}(a, bu)^\sigma - f(\tilde{h}(b, a), 1)u^\sigma \qquad \text{by B2}$$
$$= -\tilde{h}(a, bu)^\sigma + f(\tilde{h}(a, b)^\sigma, 1)u^\sigma = 0$$

for all $b \in X$ and thus $au \in V$. By 6.8, it follows that $V = X$. Thus

(6.41) $$\hat{h}(a, b) = \omega h(a, b)$$

for all $a, b \in X$.

Now let

$$\tilde{\theta}(a, w) = \hat{\theta}(a, w) - \omega \theta(a, w)$$

for all $a \in X$ and all $w \in L$ and let

$$W = \{a \in X \mid \hat{\theta}(a, w) \equiv \omega \theta(a, w) \pmod{\langle w \rangle} \text{ for all } w \in L\}.$$

Then $e \in W$ by 6.39 and W is closed under addition by C3 and 6.41. If $a \in W$, then for all $v, w \in L$, $\tilde{\theta}(a, w^\sigma)^\sigma \in \langle w \rangle$ and $\tilde{\theta}(a, v) \in \langle v \rangle$; hence

$$f(w, v^\sigma)\tilde{\theta}(a, v)^\sigma = f(\tilde{\theta}(a, v), w^\sigma)v^\sigma$$

(by 1.4) and thus

$$\tilde{\theta}(av, w) \in \langle w \rangle$$

by C4. Therefore $Wv \subset W$ for all $v \in L$. By 6.8 again, it follows that $W = X$. Hence

$$\hat{\theta}(a, w) \equiv \omega\theta(a, w) \pmod{\langle w \rangle}$$

for all $a \in X$ and all $w \in L$. Thus Ξ is equivalent to $\hat{\Xi}$.   □

We can now finish the proof of 3.2, the main result of this chapter, which we restate here:

Theorem 6.42. *Let*

$$\Xi = (K, L, q, 1, X, \cdot, h, \theta)$$

be a regular exceptional quadrangular algebra (where "regular" and "excep-tional" are as defined in 1.27 and 1.28). Then (K, L, q) is a quadratic space of type E_6, E_7 or E_8 and Ξ is uniquely determined by the pointed quadratic space $(K, L, q, 1)$.

Proof. By 4.2, we can assume that Ξ is δ-standard (for some $\delta \in L^*$), so all the results of this chapter apply. By 6.31, we know that the quadratic space (K, L, q) is of type E_6, E_7 or E_8. Let

$$\hat{\Xi} = (\hat{K}, \hat{L}, \hat{q}, \hat{1}, \hat{X}, \hat{\cdot}, \hat{h}, \hat{\theta})$$

be a quadrangular algebra with

$$(\hat{K}, \hat{L}, \hat{q}, \hat{1}) = (K, L, q, 1).$$

Let X and $\hat{X}$ have the structure of right $C(q, 1)$-modules as described in 2.23. By 2.26.i, 6.33 and 6.34, X and $\hat{X}$ are both irreducible right $C(q, 1)$-modules. By 2.26.ii, therefore, we can assume that

$$(\hat{K}, \hat{L}, \hat{q}, \hat{1}, \hat{X}, \hat{\cdot}) = (K, L, q, 1, X, \cdot).$$

By 1.26 and 6.38, it then follows that $\hat{\Xi}$ is isomorphic to Ξ.   □

Chapter Seven

Defective Quadrangular Algebras

We continue to assume that

$$\Xi = (K, L, q, 1, X, \cdot, h, \theta)$$

is a δ-standard quadrangular algebra (for some $\delta \in L$) and we continue to let f, σ, g, ϕ and π be as in 1.17. In this chapter, we assume in addition that the defect $L^\perp$ of (K, L, q) (as defined in 2.3) is non-trivial. Our goal is to show that (K, L, q) must be a quadratic space of type F_4 (as defined in 2.15) and that Ξ is uniquely determined by the pointed quadratic space $(K, L, q, 1)$.

Let $R = L^\perp$. eBy 2.4, we have $\operatorname{char}(K) = 2$. By 4.1.ii, we have $f(1, \delta) = 1$. By 1.2, therefore, $\delta^\sigma = \delta + 1$. Thus, in particular, $\sigma \neq 1$. This means that Ξ is proper (as defined in 1.27).

Proposition 7.1. $h(a, b) \in R$ for all $a, b \in X$.

Proof. Let w be a non-zero element of R. By B3,

$$f(h(aw, b), 1) = f(h(a, b), w) \in f(L, R) = 0$$

for all $a, b \in X$. By 3.5, it follows that

$$f(h(a, b), 1) = 0$$

for all $a, b \in X$. Hence

$$f(h(a, b), v) = f(h(av, b), 1) = 0$$

for all $a, b \in X$ and all $v \in L$, again by B3. $\qquad\square$

Proposition 7.2. $f(\pi(a), 1) = g(a, b) = 0$ for all $a, b \in X$.

Proof. By 4.3 and 7.1, the map g is identically zero. By 3.16, it follows that $f(\pi(a), 1) = 0$ for all $a \in X$. $\qquad\square$

Proposition 7.3. $h(a, au) = 0$ for all $a \in X$ and all $u \in L$.

Proof. Let $a \in X$ and $u \in L$. Then $h(a, au) = g(a, a)u$ by 3.15 and $g(a, a) = 0$ by 7.2. $\qquad\square$

Proposition 7.4. Let $a \in X$ and $u, v \in L$. Then

(i) $f(\theta(a, u), u) = 0$;
(ii) $f(\theta(a, u), v) = f(\theta(a, v), u)$; and
(iii) $\theta(a, \theta(a, v)) = q(\pi(a))v$.

Proof. By 7.2, $f(\pi(a), 1) = 0$. Thus (i) holds by 3.19.i, (ii) holds by 3.19.iii and (iii) holds by 4.21. □

Proposition 7.5. $\phi(a, u) = 0$ *for all* $a \in X$ *and all* $u \in R$.

Proof. Choose $a \in X$. If $u \in R$, then $u^* = u$, where u^* is as in 4.11, and hence $\phi(a, u) = 0$ by 4.12 and 7.4.i. □

Proposition 7.6. $\langle \pi(a), \theta(a, \delta) \rangle \subset \langle 1, \delta \rangle^\perp$ *and*

$$f(\theta(a, \delta), \pi(a)) = q(\pi(a))$$

for all $a \in X$.

Proof. Choose $a \in X$. We have $f(\pi(a), 1) = 0$ by 7.2, $f(\theta(a, \delta), \delta) = 0$ and

$$f(\theta(a, \delta), 1) = f(\pi(a), \delta)$$

by 7.4.i,ii and $f(\pi(a), \delta) = 0$ by 4.1. Thus

$$\langle \pi(a), \theta(a, \delta) \rangle \subset \langle 1, \delta \rangle^\perp.$$

We have $\theta(a, \pi(a)) = q(\pi(a)) \cdot 1$ by 7.4.iii, and hence

$$f(\theta(a, \delta), \pi(a)) = f(\theta(a, \pi(a)), \delta)$$
$$= q(\pi(a))f(1, \delta) = q(\pi(a))$$

by 7.4.ii. □

Let

(7.7) $$B_a = \{1, \delta, \pi(a), \theta(a, \delta)\}$$

and let

(7.8) $$W_a = \langle B_a \rangle$$

for each $a \in X^*$.

Proposition 7.9. $L = W_a \oplus W_a^\perp$, B_a *is a basis of* W_a *(so* $\dim_K W_a = 4$*),* $\theta(a, W_a) \subset W_a$ *and* $\theta(a, W_a^\perp) \subset W_a^\perp$ *for all* $a \in X^*$.

Proof. Choose $a \in X^*$. By 4.1, f restricted to $\langle 1, \delta \rangle$ is non-degenerate. Thus by 2.7,

$$L = \langle 1, \delta \rangle \oplus \langle 1, \delta \rangle^\perp.$$

By 7.6, $\langle \pi(a), \theta(a, \delta) \rangle \subset \langle 1, \delta \rangle^\perp$ and

$$f(\theta(a, \delta), \pi(a)) = q(\pi(a)),$$

so $f(\theta(a, \delta), \pi(a)) \neq 0$ by D2. Thus $\dim_K W_a = 4$ and f restricted to W_a is non-degenerate. Hence $L = W_a \oplus W_a^\perp$. By C1 and 7.4.iii, $\theta(a, W_a) \subset W_a$. By 7.4.ii, it follows that

$$f(\theta(a, u), w) = f(\theta(a, w), u) \in f(W_a, u) = 0$$

for all $w \in W_a$ and all $u \in W_a^\perp$. Therefore $\theta(a, W_a^\perp) \subset W_a^\perp$. □

Proposition 7.10. $|K| = \infty$.

Proof. This holds by 2.11 and 7.9. □

Proposition 7.11. $\theta(a, u) = h(a, a\delta u)$ for all $a \in X$ and all $u \in W_a^\perp$.

Proof. Choose $a \in X$ and $u \in W_a^\perp$. By B1 and 3.12, we can assume that $a \in X^*$. By 7.9, $\theta(a, u) \in W_a^\perp$. By 3.22 (and 7.10),

$$a\delta\theta(a, u) + a\delta\pi(a)u = ah(a, a\delta u) + ah(a, a\delta)u.$$

By 7.3, $h(a, a\delta) = 0$. By 3.8,

$$a\delta\theta(a, u) = a\theta(a, u)\delta^\sigma$$

(since $\theta(a, u) \in W_a^\perp \subset \langle 1, \delta \rangle^\perp$) and

$$a\delta\pi(a)u = a\pi(a)\delta^\sigma u = a\pi(a)u\delta$$

(since $\pi(a)$ and u both lie in $\langle 1, \delta \rangle^\perp$). We conclude that

$$
\begin{aligned}
ah(a, a\delta u) &= a\delta\theta(a, u) + a\delta\pi(a)u \\
&= a\theta(a, u)\delta^\sigma + a\pi(a)u\delta \\
&= a\theta(a, u) + \big(a\theta(a, u) + a\pi(a)u\big)\delta &&\text{by A1} \\
&= a\theta(a, u) &&\text{by D1.}
\end{aligned}
$$

Therefore $\theta(a, u) = h(a, a\delta u)$ by 3.4. □

Proposition 7.12. $W_a^\perp = R$ for all $a \in X^*$.

Proof. Choose $a \in X^*$. By D2, $q(\pi(a)) \neq 0$. We have

$$R = L^\perp \subset W_a^\perp.$$

Let $u \in W_a^\perp$ and let $w = u/q(\pi(a))$. Then $\theta(a, w) \in W_a^\perp$ by 7.9,

$$\theta(a, \theta(a, w)) = u$$

by C1 and 7.4.iii, and hence

$$u = \theta(a, \theta(a, w)) = h(a, a\delta\theta(a, w))$$

by 7.11. By 7.1, we conclude that $u \in R$. Thus $W_a^\perp \subset R$. □

Note that we do not claim (nor is it true) that W_a is independent of the choice of $a \in X^*$.

Corollary 7.13. $L = W_a \oplus R$ and

$$\theta(a, v) = h(a, a\delta v) \in R$$

for all $a \in X$ and all $v \in R$.

Proof. This holds by 7.9, 7.11 and 7.12 if $a \neq 0$ and by B1 and 3.12 if $a = 0$. □

Proposition 7.14. $auu = aq(u)$, $auv = avu$, $a\theta(a, u)v = av\theta(a, v)$ for all $a \in X$ and all $u, v \in R$.

Proof. Choose $a \in X$ and $u, v \in R$. Then $auu = aq(u)$ by A3 and $auv = avu$ by 3.8. Since $\theta(a, v) \in R$ by 7.13, it follows that $a\theta(a, u)v = av\theta(a, v)$. $\square$

Proposition 7.15. $h(a, b) = h(b, a)$ and $h(aw, b) = h(a, bw)$ for all $a, b \in X$ and all $w \in L$.

Proof. Choose $a, b \in X$ and $w \in L$. Then $h(a, b) = h(b, a)$ by 3.6 and 7.1 and thus $h(aw, b) = h(b, aw) = h(a, bw)$ for all $w \in L$ by B2 and 7.1. $\square$

Proposition 7.16. Let $a \in X$ and $u \in R$. Then

$$\theta(au, w) = \theta(a, w^\sigma)^\sigma q(u)$$

for all $w \in L$. In particular,

$$\pi(au) = \pi(a)q(u)$$

and, if $w \in R$,

$$\theta(au, w) = \theta(a, w)q(u).$$

Proof. Choose $w \in L$. We have $\theta(a, u) \in R$ by 7.13 (since $u \in R$) and hence

$$\theta(au, w) = \theta(a, w^\sigma)^\sigma q(u)$$

by C4 and 7.5. By 7.6, $\pi(a)$ is fixed by σ and by 7.13, both w and $\theta(a, w)$ are fixed by σ if $w \in R$. $\square$

Proposition 7.17. Let $C(q, 1)$ be as in 2.21 and let X be endowed with the unique right $C(q, 1)$-module structure described in 2.23. Then

$$aC(q, 1) = X$$

for all $a \in X^*$.

Proof. Choose $a \in X^*$, $b \in X$ and $v \in R^*$ and let $M = aC(q, 1)$. By 7.6, $\pi(a)$ and $\theta(a, \delta)$ are fixed by σ. Thus $bv\theta(a, \delta) = b\theta(a, \delta)v$ and

$$bv\pi(a)\delta = b\pi(a)v\delta = b\pi(a)\delta^\sigma v$$
$$= b\pi(a)\delta v + b\pi(a)v$$

by 3.8, so

$$bv\theta(a, \delta) + bv\pi(a)\delta = \big(b\theta(a, \delta) + b\pi(a)\delta\big)v + b\pi(a)v$$

by A1. By two applications of 3.22 (and 7.10),

$$bv\theta(a, \delta) + bv\pi(a)\delta = ah(a, bv)\delta + ah(a, bv\delta) \in M$$

and

$$\big(b\theta(a, \delta) + b\pi(a)\delta\big)v = \big(ah(a, b)\delta + ah(a, b\delta)\big)v \in M.$$

Therefore also $b\pi(a)v \in M$. By D2, $\pi(a) \neq 0$. By A3, it follows that

$$b \in Mv^{-1}\pi(a)^{-1} \subset M.$$

Hence $M = X$. $\square$

Proposition 7.18. $auvw \in aR$ for all $a \in X$ and all $u, v, w \in R$.

Proof. Let $a, b \in X$ and $u, v, w \in R^*$. Then

$$
\begin{aligned}
auh(au, b)v + auh(au, bv) &= b\theta(au, v) + b\pi(au)v && \text{by 3.22} \\
&= \big(b\theta(a, v) + b\pi(a)v\big)q(u) && \text{by 7.16} \\
&= \big(ah(a, b)v + ah(a, bv)\big)q(u) && \text{by 3.22.}
\end{aligned}
$$

Moving terms from one side to the other and applying A1, we obtain

$$
\begin{aligned}
(7.19) \quad auh(au, b)v + ah(a, bv)q(u) &= auh(au, bv) + ah(a, b)vq(u) \\
&= \big(ah(au, bv) + ah(a, bv)u\big)u \\
&\quad + \big(ah(a, bv) + ah(a, b)v\big)q(u)
\end{aligned}
$$

since $auh(au, bv) = ah(au, bv)u$ and $ah(a, bv)uu = ah(a, bv)q(u)$ (by 7.1 and 7.14). We have

$$ b\pi(a)v = bv\pi(a) $$

by 3.8 (since $\pi(a)^\sigma = \pi(a)$ by 7.6). Thus

$$
\begin{aligned}
ah(au, bv) + ah(a, bv)u &= ah(a, bvu) + ah(a, bv)u && \text{by 7.15} \\
&= bv\theta(a, u) + bv\pi(a)u && \text{by 3.22} \\
&= b\theta(a, u)v + b\pi(a)vu && \text{by 7.14.}
\end{aligned}
$$

Also

$$ ah(a, bv) + ah(a, b)v = b\theta(a, v) + b\pi(a)v, $$

again by 3.22. By 7.19, therefore,

$$
\begin{aligned}
auh(au, b)v + ah(a, bv)q(u) &= \big(b\theta(a, u)v + b\pi(a)vu\big)u \\
&\quad + \big(b\theta(a, v) + b\pi(a)v\big)q(u).
\end{aligned}
$$

Since $b\pi(a)vuu = b\pi(a)vq(u)$ and $b\theta(a, v)uu = b\theta(a, v)q(u)$ by 7.14, we conclude that

$$ auh(au, b)v + ah(a, bv)q(u) = \big(b\theta(a, u)v + b\theta(a, v)u\big)u. $$

Replacing b by bu, we obtain

$$ auh(au, bu)v + ah(a, buv)q(u) = \big(bu\theta(a, u)v + bu\theta(a, v)u\big)u. $$

Since $h(au, bu) = h(a, buu) = h(a, b)q(u)$ by B1, 7.14 and 7.15, it follows (again by 7.14) that

$$ \big(auh(a, b)v + ah(a, buv)\big)q(u) = \big(b\theta(a, u)v + b\theta(a, v)u\big)q(u). $$

Next we cancel $q(u)$ from both sides and replace b by $a\delta w$ to obtain

$$ (7.20) \quad auh(a, a\delta w)v + ah(a, a\delta wuv) = a\delta w\theta(a, u)v + a\delta w\theta(a, v)u. $$

By A1 and 3.8, we have

$$ a\delta w\theta(a, u)v + a\delta w\theta(a, v)u = \big(a\theta(a, u)v + a\theta(a, v)u\big)\delta w. $$

By D1,

$$a\theta(a,u)v + a\theta(a,v)u = a\pi(a)uv + a\pi(a)vu$$

and

$$a\pi(a)uv + a\pi(a)vu = 0$$

by 7.14. By 7.20, therefore,

$$auh(a,a\delta w)v = ah(a,a\delta wuv),$$

so

(7.21) $$au\theta(a,w)v = ah(a,a\delta wuv)$$

by 7.13. Now suppose that $a \neq 0$, so $q(\pi(a)) \neq 0$ by D2, and let $\tilde{w} = \theta(a, w/q(\pi(a)))$. By C1 and 7.4.iii,

$$w = \theta(a,\tilde{w}).$$

Replacing w by $\tilde{w}$ in 7.21, we therefore obtain

$$auwv = ah(a,a\delta\tilde{w}uv).$$

By 7.1, $h(a,a\delta\tilde{w}uv) \in R$. Thus $auvw \in aR$. If $a = 0$, then $auvw = 0 \in aR$ by A1. □

Notation 7.22. Let ρ be a non-zero element of R; let

(7.23) $$\beta = q(\rho);$$

let

$$(a,u) \mapsto a * u$$

denote the map from $X \times R$ to X given by

(7.24) $$a * u = au\rho^{-1}$$

for all $a \in X$ and all $u \in R$; and let τ denote the map from R to K given by

(7.25) $$\tau(u) = q(u)/q(\rho)$$

for all $u \in R$.

Note that the map τ is additive (since $R = L^{\perp}$) and injective (since q is anisotropic).

Proposition 7.26. *Let $u, v \in R$. Then there exists a unique element $w \in R$ such that*

$$\tau(u)\tau(v) = \tau(w).$$

Moreover, $au\rho^{-1}v = aw$ for all $a \in X$.

Proof. Choose $a \in X^*$ (so $\pi(a) \neq 0$ by D2) and $u, v \in R$. By 7.18, there exists an element $w \in R$ (which depends on a as well as on u and v and the choice of ρ) such that

$$(7.27) \qquad\qquad au\rho^{-1}v = aw.$$

Then $\pi(au\rho^{-1}v) = \pi(aw)$. By 7.16, it follows that

$$\pi(a)q(u)q(\rho^{-1})q(v) = \pi(a)q(w),$$

hence $q(u)q(\rho^{-1})q(v) = q(w)$, and therefore (since $q(\rho^{-1}) = q(\rho)^{-1}$ by 1.3)

$$(7.28) \qquad\qquad \tau(u)\tau(v) = \tau(w).$$

Since τ is injective, the element w is uniquely determined by the identity 7.28. Therefore the identity 7.27 holds for *all* $a \in X$. □

Proposition 7.29. *Let ρ, $*$ and τ be as in 7.22. There is a unique multiplication on R (which we will denote by juxtaposition) with respect to which the map τ is multiplicative. Moreover:*

(i) R endowed with this multiplication is a (commutative) field with multiplicative identity ρ;

(ii) τ is an injective homomorphism (of fields) from R into K; and

(iii) X is a vector space over R with scalar multiplication $*$.

Proof. By 7.26, there is a unique multiplication on R with respect to which τ is multiplicative. The map τ (which we already knew to be additive and injective) is thus an isomorphism from R (endowed with this multiplication) to a subring of K and $\tau(\rho) = 1$. If $u \in R^*$, then

$$\tau(u)\tau(q(\rho)u^{-1}) = q(u)/q(\rho) \cdot q(\rho)^2 q(u^{-1})/q(\rho)$$
$$= q(u)/q(\rho) \cdot q(\rho)/q(u) = 1.$$

It follows that $\tau(R)$ is a field (which is commutative since K is commutative) with multiplicative identity ρ. Thus (i) and (ii) hold.

Choose $u, v \in R$ and let $w = uv$. By 7.26,

$$a * w = aw\rho^{-1} = au\rho^{-1}v\rho^{-1} = (a * u) * v$$

for all $a \in X$. By A1 and A2, the triple $(R, X, *)$ satisfies the other axioms of a vector space. Thus (iii) holds. □

Notation 7.30. Let ρ, $*$ and τ be as in 7.22 and let

$$(7.31) \qquad\qquad F = \tau(R).$$

By 7.29, F is a subfield of K. Since $\tau(t\rho) = q(t\rho)/q(\rho) = t^2$ for all $t \in K$, we have

$$(7.32) \qquad\qquad K^2 \subset F \subset K.$$

Let $\circ$ denote the map from $X \times F$ to X given by

$$(7.33) \qquad\qquad a \circ s = a * \tau^{-1}(s)$$

for all $a \in X$ and all $s \in F$. By 7.29 again, X is a vector space over F with respect to $\circ$.

Note that the original scalar multiplication from $X \times K$ to X can be recovered from $\circ$ by restriction of scalars:

Proposition 7.34. $at = a \circ t^2$ for all $a \in X$ and all $t \in K$.

Proof. Choose $a \in X$ and $t \in K$. Then
$$a \circ t^2 = a * \tau^{-1}(t^2) = a * (t\rho) = a(t\rho)\rho^{-1} = at$$
by A1, A3 and 7.30. $\square$

Proposition 7.35. $av \circ s = (a \circ s)v$ for all $a \in X$, all $v \in L$ and all $s \in F$.

Proof. Choose $a \in X$, $v \in L$ and $s \in F$ and let $u = \tau^{-1}(s)$. Then
$$av \circ s = avu\rho^{-1} = auv^\sigma \rho^{-1} = au\rho^{-1}v = (a \circ s)v$$
by 7.30 and two applications of 3.8. $\square$

Proposition 7.36. Let $a \in X^*$. Then $q(\pi(a)) = \tau(\theta(a,\rho)) \in F^*$ and
$$a\pi(a) = a \circ q(\pi(a)).$$

Proof. By 7.13, $\theta(a,\rho) \in \theta(a,R) \subset R$. Therefore
$$a\pi(a)\rho = a\theta(a,\rho) = a\theta(a,\rho)\rho^{-1}\rho$$
$$= \big(a * \theta(a,\rho)\big)\rho = \big(a \circ \tau(\theta(a,\rho))\big)\rho,$$
and hence
$$a\pi(a) = a \circ \tau(\theta(a,\rho))$$
by A3, D1 and 7.30 and
$$q(\pi(a)) = q(\theta(a,\rho))/q(\rho) = \tau(\theta(a,\rho)) \in F$$
by 4.22 and 7.31. We have $q(\pi(a)) \neq 0$ by D2. $\square$

Notation 7.37. Choose $e \in X^*$, let
$$(7.38) \qquad \alpha = q(\pi(e))$$
(so $\alpha \in F^*$ by 7.36) and let $W = W_e = \langle 1, \delta, \pi(e), \theta(e,\delta) \rangle$ (as defined in 7.7 and 7.8). Thus
$$(7.39) \qquad L = W \oplus R$$
by 7.13.

Note that by 7.32, we can apply 2.14 to the pair F and K.

Proposition 7.40. Let
$$p(x) = x^2 + x + q(\delta) \in K[x],$$
let E denote the splitting field of $p(x)$ over K and let F be as in 7.31. Then E/K is a separable quadratic extension and
$$(K,L,q) \cong (K, E \oplus E, N \perp \alpha N) \perp (K,F,\beta q_F),$$
where N denotes the norm of the extension E/K, α is as in 7.38, β is as in 7.23 and (K,F,q_F) is as in 2.14. More precisely,
$$(K, \langle 1,\delta \rangle, q|_{\langle 1,\delta \rangle}) \cong (K,E,N)$$
and
$$(K, \langle 1,\delta \rangle^\perp, q|_{\langle 1,\delta \rangle^\perp}) \cong (K,E,\alpha N) \perp (K,F,\beta q_F).$$

Proof. Setting $u = 1$ and $v = \delta$ in 2.10, we conclude that E/K is a separable quadratic extension and that

$$(K, \langle 1, \delta \rangle, q|_{\langle 1, \delta \rangle}) \cong (K, E, N).$$

By 4.22, $q(\theta(e, \delta)) = \alpha q(\delta)$ (where e is as in 7.37), and by 7.6,

$$f(\theta(e, \delta), \pi(e)) = \alpha.$$

Setting $u = \pi(e)$ and $v = \theta(e, \delta)$ in 2.10, we thus have

$$(K, \langle \pi(e), \theta(e, \delta) \rangle, q|_{\langle \pi(e), \theta(e, \delta) \rangle}) \cong (K, E, \alpha N).$$

Thus by 7.6 and 7.37, we have

$$\langle 1, \delta \rangle^{\perp} = \langle \pi(e), \theta(e, \delta) \rangle \oplus R$$

and

$$(K, W, q|_W) \cong (K, E, N) \perp (K, E, \alpha N).$$

By 7.39, we have

$$(K, L, q) \cong (K, W, q|_W) \perp (K, R, q|_R).$$

By 2.14 and 7.25,

$$\beta q_F(\tau(u)) = \beta \tau(u) = q(u)$$

for all $u \in R$. It follows that the map τ is a K-linear isomorphism (of quadratic spaces) from $(K, R, q|_R)$ to $(K, F, \beta q_F)$. □

Proposition 7.41. *The quadratic space (K, L, q) is of type F_4 (as defined in 2.15).*

Proof. By 7.37, $\alpha \in F$. The claim holds, therefore, by 7.32 and 7.40. □

Proposition 7.42. *Let $p(x)$, F, E and N be as in 7.40, let γ be a root of $p(x)$ in E, let*

$$p_0(x) = x^2 + x + q(\delta)^2,$$

let $D = F(\gamma^2)$ and let β be as in 7.23. Then the following hold:

(i) *$p_0(x) \in F[x]$, D is the splitting field of $p_0(x)$ over F, D/F is a separable quadratic extension and the unique non-trivial element of $\mathrm{Gal}(E/K)$ restricts to the unique non-trivial element of $\mathrm{Gal}(D/F)$.*

(ii) *$N(E) \cap F(\beta) = K^2 \cdot N(D(\beta))$.*

Proof. Let $w \mapsto \bar{w}$ denote the non-trivial element of $\mathrm{Gal}(E/K)$. Thus $\bar{\gamma} = \gamma + 1$. By 7.32, $p_0(x) \in F[x]$. Since $\gamma^2 = \gamma + q(\delta)$, we have $\gamma^4 = \gamma^2 + q(\delta)^2$, i.e. $p_0(\gamma^2) = 0$. Since $\bar{\gamma}^2 = \gamma^2 + 1 \neq \gamma^2$, the restriction of $w \mapsto \bar{w}$ to D is a non-trivial element in $\mathrm{Gal}(D/F)$. Thus (i) holds.

By 7.32, $K^2 \subset N(E) \cap F$ and $\beta^2 \in F$. Since the fixed field of the map $w \mapsto \bar{w}$ in D is F, we have $N(u) = u\bar{u} \in F$ and

$$u\bar{v} + v\bar{u} \in F$$

for all $u, v \in D$. Therefore

$$N(u + \beta v) = (u + \beta v)(\bar{u} + \beta \bar{v})$$
$$= N(u) + \beta^2 N(v) + \beta(u\bar{v} + v\bar{u}) \in F(\beta)$$

for all $u, v \in D$. Since $D(\beta) \subset E$, we conclude that

$$K^2 \cdot N(D(\beta)) \subset N(E) \cap F(\beta).$$

Choose $u \in E$. Since $\bar{\gamma}^2 \neq \gamma^2$, we have $E = K(\gamma^2)$. There thus exist $r, t \in K$ such that $u = r + t\gamma^2$. Suppose now that $N(u) \in F(\beta)$. We claim that under this assumption, $N(u) \in K^2 \cdot N(D(\beta))$. If $t = 0$, then $N(u) = r^2 \in K^2$. We can assume, therefore, that $t \neq 0$. Let $z = r/t$. We have

$$N(z + \gamma^2) = (z + \gamma^2)(z + \bar{\gamma}^2)$$
$$= (z + \gamma^2)(z + \gamma^2 + 1) = q(\delta)^2 + z^2 + z$$

and thus

$$N(u) = N(r + t\gamma^2) = t^2 N(z + \gamma^2) = t^2(z^2 + z + q(\delta)^2).$$

Since $N(u) \in F(\beta)$ (by assumption) and t^2 and $(z + q(\delta))^2$ lie in $K^2 \subset F \subset F(\beta)$, it follows that $z \in F(\beta)$. Therefore $z + \gamma^2 \in D(\beta)$. We conclude that

$$N(u) = t^2 N(z + \gamma^2) \in K^2 \cdot N(D(\beta)).$$

Thus (ii) holds. □

Proposition 7.43. $\beta \notin F$.

Proof. By 7.40, the quadratic space

$$(K, E, N) \perp (K, F, \beta q_F)$$

is anisotropic. Since $\beta^2 \in K^2 \subset N(E)$, it follows that $\beta \notin q_F(F)$. By 2.14, $q_F(F) = F$. □

We observe that by 7.32 and 7.43, the field K is imperfect.

For the next result, recall that $q(\pi(X)) \subset F$ by 7.36; $h(X, X) \subset R$ by 7.1; and R is the domain of τ (by 7.22).

Proposition 7.44. Let $Q(a) = q(\pi(a))$ for all $a \in X$ and let

(7.45) $$H(a, b) = \tau(h(a, b\rho))$$

for all $a, b \in X$. Then Q is an anisotropic quadratic form on X (as a vector space over F with respect to $\circ$) and H is the corresponding bilinear form.

Proof. Choose $a, b \in X$ and $s \in F$ and let $u = \tau^{-1}(s)$. Then

$$Q(a \circ s) = Q(au\rho^{-1}) \qquad \text{by 7.30}$$
$$= q(\pi(a)q(u)/q(\rho)) \qquad \text{by 7.16}$$
$$= Q(a)\tau(u)^2 = Q(a)s^2.$$

By C3 and 7.2, we have

$$h(a, bv) = \theta(a + b, v) + \theta(a, v) + \theta(b, v)$$

for all $v \in L$. By 7.13, $\theta(X, v) \subset R$ for all $v \in R$. Therefore

(7.46)
$$\begin{aligned} q(h(a, bv)) &= q(\theta(a + b, v)) + q(\theta(a, v)) + q(\theta(b, v)) \\ &= \big(Q(a + b) + Q(a) + Q(b)\big)q(v) \end{aligned}$$

for all $v \in R$ by 4.22. Setting $v = \rho$, we obtain

(7.47)
$$H(a, b) = Q(a + b) + Q(a) + Q(b)$$

by 7.25. Therefore

$$q(h(a, bv)) = H(a, b)q(v)$$

for all $v \in R$ by 7.46. Dividing both sides by $q(\rho)$, we obtain

(7.48)
$$\tau(h(a, bv)) = H(a, b)\tau(v)$$

for all $v \in R$ by 7.25. Thus

$$\begin{aligned} H(a, b \circ s) &= H(a, bu\rho^{-1}) && \text{by 7.30} \\ &= \tau(h(a, bu\rho^{-1}\rho) && \text{by 7.45} \\ &= \tau(h(a, bu)) && \text{by A3} \\ &= H(a, b)\tau(u) && \text{by 7.48} \\ &= H(a, b)s. \end{aligned}$$

Since H is symmetric (by 7.47) and bi-additive (by B1), it is therefore bilinear (over F). $\qquad \square$

Proposition 7.49. *Let H be as in 7.45, let α be as in 7.38 and let β be as in 7.23 (so α and β are both non-zero). Then $H(e, e\delta) = \alpha$,*

$$H(e\rho, e\delta\rho) = \alpha\beta^2$$

and $H(a, b) = 0$ for all $(a, b) \in \{e, e\delta\} \times \{e\rho, e\delta\rho\}$.

Proof. We have

$$\begin{aligned} H(e, e\delta) &= \tau(h(e, e\delta\rho)) && \text{by 7.45} \\ &= \tau(\theta(e, \rho)) && \text{by 7.13} \\ &= q(\pi(e)) = \alpha && \text{by 7.36 and 7.38} \end{aligned}$$

and hence

$$\begin{aligned} H(e\rho, e\delta\rho) &= \tau(h(e\rho, e\delta\rho^2)) && \text{by 7.45} \\ &= \tau(h(e\rho, e\delta)\beta) && \text{by B1, 7.14 and 7.23} \\ &= \tau(h(e, e\delta\rho))\beta^2 && \text{by 7.15 and 7.25} \\ &= H(e, e\delta)\beta^2 = \alpha\beta^2 && \text{by 7.45.} \end{aligned}$$

Now let $a \in \{e, e\delta\}$ and $b \in \{e\rho, e\delta\rho\}$. Then $b\rho \in \{e, e\delta\}\beta$ by 7.14 and thus

$$h(a, b\rho) = 0$$

by B1, 7.3 and 7.15. Hence $H(a, b) = 0$. $\qquad \square$

Proposition 7.50. *Let*

$$B_X = \{e, e\delta, e\rho, e\delta\rho\}.$$

Then B_X is a basis of X (as a vector space over F with respect to $\circ$).

Proof. By 7.44 and 7.49, the set B_X is linearly independent over F. Let Y denote the subspace of X spanned by B_X over F. Suppose that

(7.51) $$B_X B_e \cup B_X R \subset Y,$$

where B_e is as in 7.7. By A1 and 7.35, it follows that

$$Y B_e + Y R \subset Y.$$

By 7.39, $B_e \cup R$ spans L over K. Hence by A1 and 7.34,

$$YL \subset Y.$$

By 7.17, this implies that $Y = X$. It thus suffices to show that 7.51 holds. We have

(7.52)
$$
\begin{aligned}
a\rho\rho &= aq(\rho) && \text{by 7.14} \\
&= a\beta && \text{by 7.23} \\
&= a \circ \beta^2 && \text{by 7.34}
\end{aligned}
$$

for all $a \in X$ and

(7.53)
$$
\begin{aligned}
au &= a\rho\rho^{-1}u && \text{by A3} \\
&= a\rho u\rho^{-1} && \text{by 7.14} \\
&= a\rho * u && \text{by 7.24} \\
&= a\rho \circ \tau(u) && \text{by 7.33}
\end{aligned}
$$

for all $a \in X$ and all $u \in R$. Thus

$$B_X R \subset Y.$$

By A2, $B_X \cdot 1 \subset Y$. Next, we have

$$
\begin{aligned}
e\delta\delta &= e\delta + eq(\delta) && \text{by 3.10} \\
&= e\delta + e \circ q(\delta)^2 && \text{by 7.34,}
\end{aligned}
$$

as well as

$$
\begin{aligned}
e\rho\delta &= e\delta^\sigma \rho && \text{by 3.8} \\
&= e\rho + e\delta\rho && \text{by 1.2}
\end{aligned}
$$

and

$$
\begin{aligned}
e\delta\rho\delta &= e\rho q(\delta) && \text{by 3.9} \\
&= e\rho \circ q(\delta)^2 && \text{by 7.34.}
\end{aligned}
$$

Hence $B_X\delta \subset Y$. By 7.36 and 7.38, $e\pi(e) = e \circ \alpha$. By 7.6, $\pi(e)^\sigma = \pi(e)$. Thus

$$\begin{aligned}
e\delta\pi(e) &= e\pi(e)\delta^\sigma & \text{by 3.8} \\
&= (e \circ \alpha)\delta^\sigma \\
&= (e + e\delta) \circ \alpha & \text{by 7.35,}
\end{aligned}$$

as well as

(7.54)
$$\begin{aligned}
e\rho\pi(e) &= e\pi(e)\rho & \text{by 3.8} \\
&= (e\rho) \circ \alpha & \text{by 7.35}
\end{aligned}$$

and

$$\begin{aligned}
e\delta\rho\pi(e) &= e\pi(e)\delta^\sigma\rho & \text{by 3.8} \\
&= (e\rho + e\delta\rho) \circ \alpha & \text{by 7.35.}
\end{aligned}$$

Hence $B_X\pi(e) \subset Y$. By 7.6, $\theta(e,\delta)^\sigma = \theta(e,\delta)$. Continuing to keep in mind that $e\pi(e) = e \circ \alpha$, we thus have

$$\begin{aligned}
e\theta(e,\delta) &= e\pi(e)\delta & \text{by D1} \\
&= (e\delta) \circ \alpha & \text{by 7.35,}
\end{aligned}$$

as well as

$$\begin{aligned}
e\delta\theta(e,\delta) &= e\theta(e,\delta)\delta^\sigma & \text{by 3.8} \\
&= e\pi(e)\delta\delta^\sigma = e\pi(e)q(\delta) & \text{by A3 and D1} \\
&= e \circ (\alpha q(\delta)^2) & \text{by 7.34}
\end{aligned}$$

and

$$\begin{aligned}
e\rho\theta(e,\delta) &= e\theta(e,\delta)\rho & \text{by 3.8} \\
&= e\pi(e)\delta\rho & \text{by D1} \\
&= (e\delta\rho) \circ \alpha & \text{by 7.35.}
\end{aligned}$$

Finally,

$$\begin{aligned}
e\delta\rho\theta(e,\delta) &= e\theta(e,\delta)\delta^\sigma\rho & \text{by 3.8} \\
&= e\pi(e)\delta\delta^\sigma\rho & \text{by D1} \\
&= e\pi(e)\rho \circ q(\delta)^2 & \text{by A3 and 7.34} \\
&= e\rho \circ (\alpha q(\delta)^2) & \text{by 7.35.}
\end{aligned}$$

Thus also $B_X\theta(e,\delta) \subset Y$. We conclude that 7.51 holds.   □

Proposition 7.55. $(F, X, Q) \cong (F, D \oplus D, \alpha N_D \perp \alpha\beta^2 N_D)$, where Q is as in 7.44, α is as in 7.38, β is as in 7.23, D is as in 7.42 and N_D is the norm of the extension D/F (so N_D is the restriction to D of the norm of the extension E/K by 7.42.i).

Proof. We have $\phi(e,\delta) = 0$ by 4.13. By C4 and 7.6, therefore,

$$\pi(e\delta) = \pi(e)q(\delta) + \theta(e,\delta).$$

We have

$$q(\theta(e,\delta)) = \alpha q(\delta)$$

by 4.22 and

$$f(\theta(e,\delta), \pi(e)) = \alpha$$

by 7.6. Therefore

$$\begin{aligned}
Q(e\delta) &= q(\pi(e\delta)) \\
&= q(\pi(e)q(\delta) + \theta(e,\delta)) \\
&= q(\pi(e))q(\delta)^2 + q(\theta(e,\delta)) + f(\pi(e), \theta(e,\delta))q(\delta) \\
&= Q(e)q(\delta)^2 = \alpha q(\delta)^2.
\end{aligned}$$

Hence

$$Q(e)x^2 + H(e, e\delta)x + Q(e\delta) = \alpha(x^2 + x + q(\delta)^2)$$

by 7.49. By 7.42.i, D is the splitting field of $x^2 + x + q(\delta)^2$ over F. By 2.10, therefore,

$$(F, \langle e, e\delta \rangle, Q|_{\langle e, e\delta \rangle}) \cong (F, D, \alpha N_D).$$

By 7.16, $Q(a\rho) = Q(a)q(\rho)^2 = Q(a)\beta^2$ for all $a \in X$. Therefore $Q(e\rho) = \alpha\beta^2$ and $Q(e\delta\rho) = \alpha\beta^2 q(\delta)^2$. Hence

$$Q(e\rho)x^2 + H(e\rho, e\delta\rho)x + Q(e\delta\rho) = \alpha\beta^2(x^2 + x + q(\delta)^2)$$

by 7.49. By another application of 2.10, we conclude that

$$(F, \langle e\rho, e\delta\rho \rangle, Q|_{\langle e\rho, e\delta\rho \rangle}) \cong (F, D, \alpha\beta^2 N_D).$$

By 7.49 again, $\langle e, e\delta \rangle$ and $\langle e\rho, e\delta\rho \rangle$ are perpendicular with respect to H. Thus

$$(F, X, Q) \cong (F, D \oplus D, \alpha N_D \perp \alpha\beta^2 N_D).$$

$\square$

Proposition 7.56. $q(\langle 1, \delta \rangle^\perp) \cap F = K^2 \cdot Q(X)$, where Q is as in 7.44.

Proof. By 7.6, $\pi(a) \in \langle 1, \delta \rangle^\perp$ for all $a \in X$. By 7.36, therefore,

$$Q(X) \subset q(\langle 1, \delta \rangle^\perp) \cap F.$$

By 1.1.ii and 7.32, the set $q(\langle 1, \delta \rangle^\perp) \cap F$ is closed under left multiplication by elements of K^2. Thus, in fact,

$$K^2 \cdot Q(X) \subset q(\langle 1, \delta \rangle^\perp) \cap F.$$

Let $u \in \langle 1, \delta \rangle^\perp$ and suppose that $q(u) \in F$. It will suffice now to show that $q(u) \in K^2 \cdot Q(X)$. By 7.40, there exist elements $v \in E$ and $s \in F$ such that $q(u) = \alpha N(v) + \beta s$, where N denotes the norm of the extension

E/K. Since $\alpha \in F^*$ (by 7.36) and $q(u) \in F$ (by hypothesis), it follows that $N(v) \in F(\beta)$. By 7.42.ii, therefore, there exist elements $\omega \in K$ and $x, y \in D$ such that

$$N(v) = \omega^2 N(x + \beta y)$$
$$= \omega^2 (x + \beta y)(\bar{x} + \beta \bar{y})$$
$$= \omega^2 \big(N(x) + \beta^2 N(y) + \beta(x\bar{y} + y\bar{x})\big),$$

where $w \mapsto \bar{w}$ is the unique non-trivial element of $\mathrm{Gal}(E/K)$. Since $x, y \in D$, the elements $x\bar{y} + y\bar{x}$, $N(x) = x\bar{x}$ and $N(y) = y\bar{y}$ all lie in F (i.e. in the set of fixed points of the map $w \mapsto \bar{w}$ in D). Thus (since both ω^2 and β^2 lie in F by 7.32 and $\alpha \in F$ by 7.36),

$$q(u) \equiv \alpha\omega^2 \big(N(x) + \beta^2 N(y)\big) \pmod{\beta F}$$

and

$$\alpha\omega^2 (N(x) + \beta^2 N(y)) \in F.$$

By 7.43, on the other hand, $\beta \notin F$. Since $q(u) \in F$, we conclude that

$$q(u) = \alpha\omega^2 \big(N(x) + \beta^2 N(y)\big).$$

By 7.55, $\alpha(N(x) + \beta^2 N(y)) \in Q(X)$. Therefore

$$q(u) \in K^2 \cdot Q(X).$$

$\square$

We can now finish the proof of 3.3, the main result of the this chapter, which we restate here:

Theorem 7.57. *Let*

$$\Xi = (K, L, q, 1, X, \cdot, h, \theta)$$

be a defective but proper quadrangular algebra. Then (K, L, q) is a quadratic space of type F_4 and Ξ is uniquely determined by the pointed quadratic space $(K, L, q, 1)$.

Proof. By 4.2, we can assume that Ξ is δ-standard (for some $\delta \in L^*$). By 7.41, (K, L, q) is a quadratic space of type F_4. Suppose that

$$\hat{\Xi} = (\hat{K}, \hat{L}, \hat{q}, \hat{1}, \hat{X}, \hat{\cdot}, \hat{h}, \hat{\theta})$$

is a quadrangular algebra such that

$$(\hat{K}, \hat{L}, \hat{q}, \hat{1}) = (K, L, q, 1).$$

By 4.2, we can assume that $\hat{\Xi}$ is also δ-standard. Choose $\hat{e} \in \hat{X}^*$ and let $\hat{\alpha} = q(\hat{\pi}(\hat{e}))$. We have $\hat{\pi}(\hat{e}) \in \langle 1, \delta \rangle^\perp$ by 7.6 and $\hat{\alpha} = q(\hat{\pi}(\hat{e})) \in F^*$ by 7.36. By 7.56, therefore, there exist elements $\omega \in K^*$ and $a \in X^*$ such that $\hat{\alpha} = \omega^2 q(\pi(a))$. By 7.37, e is an arbitrary element of X^*. Hence we can assume without loss of generality that $a = e$. Thus $\hat{\alpha} = \omega^2 \alpha$. By 7.9, there exists a unique K-linear map ψ_1 from L to itself which acts trivially on

$\langle 1, \delta \rangle \cup R$ and maps $\pi(e)$ to $\omega^{-1}\hat{\pi}(\hat{e})$ and $\theta(e, \delta)$ to $\omega^{-1}\hat{\theta}(\hat{e}, \delta)$. By 4.22 and 7.6, ψ_1 is a K-linear automorphism of the pointed quadratic space $(K, L, q, 1)$. By 7.50,

$$\{e, e\delta, e\rho, e\delta\rho\}$$

is a basis of X over F and

$$\{\hat{e}, \hat{e}\delta, \hat{e}\rho, \hat{e}\delta\rho\}$$

is a basis of $\hat{X}$ over F (with respect to the map $\circ$ defined in 7.30 applied to $\hat{\Xi}$). (Note that in expressions like $e\delta$ and $\hat{e}\delta$ we are using juxtaposition to denote $\cdot$ in the first case and $\hat{\ }$ in the second. We allow ourselves to use juxtaposition to denote these two maps only where it is clear whether the first factor is in X, in which case we mean $\cdot$, or in $\hat{X}$, in which case we mean $\hat{\ }$.) Let ψ_2 be the unique F-linear map from X to $\hat{X}$ which sends e to $\hat{e}$, $e\delta$ to $\hat{e}\delta$, $e\rho$ to $\hat{e}\rho$ and $e\delta\rho$ to $\hat{e}\delta\rho$. By 7.34, ψ_2 is also linear over K. By 7.34 and the calculations in the proof of 7.50, we have

$$(7.58) \qquad\qquad \psi_2(a \cdot v) = \psi_2(a) \hat{\ } \psi_1(v)$$

for all a in

$$\{e, e\delta, e\rho, e\delta\rho\}$$

and all v in

$$\{1, \delta, \pi(e), \theta(e, \delta)\} \cup R.$$

(For example,

$$
\begin{aligned}
\psi_2(e\rho \cdot \pi(a)) &= \psi_2(e\rho \circ \alpha) && \text{by 7.54} \\
&= \psi_2(e\rho) \circ \alpha \\
&= \hat{e}\rho \circ \alpha \\
&= (\hat{e}\rho \circ \hat{\alpha})\omega^{-1} && \text{by 7.34} \\
&= (\hat{e}\rho)\hat{\ }(\hat{\pi}(e)\omega^{-1}) && \text{by A1 and 7.54} \\
&= \psi_2(e\rho)\hat{\ }\psi_1(\pi(e))
\end{aligned}
$$

and

$$
\begin{aligned}
\psi_2(e\rho \cdot u) &= \psi_2(e\rho\rho \circ \tau(u)) && \text{by 7.53} \\
&= \psi_2(e \circ (\beta^2\tau(u))) && \text{by 7.52} \\
&= \psi_2(e) \circ (\beta^2\tau(u)) \\
&= \hat{e} \circ (\beta^2\tau(u)) \\
&= \hat{e}\rho\rho \circ \tau(u) && \text{by 7.52} \\
&= (\hat{e}\rho)\hat{\ }u = \psi_2(e\rho)\hat{\ }\psi_1(u) && \text{by 7.53}
\end{aligned}
$$

for all $u \in R$.) By A1, 7.35 and 7.39, it follows that the identity 7.58 holds for all $a \in X$ and all $v \in L$. By 7.45 and 7.49 (applied to both Ξ and to $\hat{\Xi}$), we have

$$\tau(\hat{h}(\psi_2(a), \psi_2(b\rho))) = \tau(\hat{h}(\psi_2(a), \psi_2(b)\hat{\cdot}\psi_1(\rho)))$$
$$= \tau(\hat{h}(\psi_2(a), \psi_2(b)\hat{\cdot}\rho))$$
$$= \omega^2\tau(h(a, b\rho))$$
$$= \tau(\omega h(a, b\rho))$$
$$= \tau(\omega\psi_1(h(a, b\rho)))$$

and thus (since $X = X\rho$ by 3.5),

$$\hat{h}(\psi_2(a), \psi_2(b)) = \omega\psi_1(h(a, b))$$

for all $a, b \in X$. By 7.4.iii and 7.13, therefore,

(7.59) $$\hat{\theta}(\hat{e}, \psi_1(v)) = \omega\psi_1(\theta(e, v))$$

for all v in

$$\{1, \delta, \pi(e), \theta(e, \delta)\} \cup R.$$

By C1, it follows that the identity 7.59 holds for all $v \in L$. By C3, C4, 4.12, 7.2 and 7.17, we conclude that

$$\hat{\theta}(\psi_2(a), \psi_1(v)) = \omega\psi_1(\theta(a, v))$$

for all $a \in X$ and all $v \in L$. Therefore (ψ_1, ψ_2) is an isomorphism from X to $\hat{X}$ as defined in 1.25. $\qquad\square$

Chapter Eight

Isotopes

In the geometrical considerations which gave rise to the notion of a quadrangular algebra, there is nothing canonical about the basepoint. To capture this idea algebraically, we introduce in this chapter the notion of *isotopic* quadrangular algebras.

Recall that by 3.14 and 4.2, it is no real restriction to assume that a quadrangular algebra is standard (as defined in 4.1) if the characteristic is different from two.

Proposition 8.1. *Let*

$$\Xi = (K, L, q, 1, X, \cdot, h, \theta)$$

be a quadrangular algebra and let σ, g *and* ϕ *be as in 1.17. Suppose that* Ξ *is standard if* $\mathrm{char}(K) \neq 2$ *(but if* $\mathrm{char}(K) = 2$, Ξ *may or may not be proper). Let* u *be an element of* L^* *and let*

$$\hat{\Xi} = (K, L, \hat{q}, \hat{1}, X, \hat{\cdot}, \hat{h}, \hat{\theta}),$$

where

$$\hat{q}(v) = q(u)^{-1} q(v),$$

$$\hat{1} = u,$$

$$a \hat{\cdot} v = (av)u^{-1},$$

$$\hat{h}(a, b) = h(a, bu)q(u)^{-1} \text{ and}$$

$$\hat{\theta}(a, v) = q(u)^{-1}\theta(a, v)$$

for all $a, b \in X$ *and all* $v \in L$. *Then* $\hat{\Xi}$ *is a quadrangular algebra which is standard if* $\mathrm{char}(K) \neq 2$, *and the maps* $\hat{g}$ *and* $\hat{\phi}$ *which occur in C3 and C4 are given by*

$$\hat{g}(a, b) = g(a, b)q(u)^{-1} \text{ and}$$

$$\hat{\phi}(a, v) = \Big(\phi(av, u^\sigma) + q(u)\phi(a, v)\Big)q(u)^{-3}$$

for all $a, b \in X$ *and all* $v \in L$.

Proof. It is clear that $\hat{\Xi}$ satisfies A1, A2, B1, C1 and C2. Let

$$\hat{f} = q(u)^{-1}f,$$

so $\hat{f}$ is the bilinear form associated with $\hat{q}$, and let $\hat{\sigma}$ denote the standard involution of the pointed quadratic space $(K, L, \hat{q}, \hat{1})$. By 1.2,

(8.2)
$$v^{\hat{\sigma}} = \hat{f}(v, \hat{1})\hat{1} - v$$
$$= \big(f(v, u)u - q(u)v\big)q(u)^{-1}$$

for all $v \in L$. Choose $a \in X$ and $v \in L^*$ and let

$$z = v^{\hat{\sigma}}/\hat{q}(v).$$

Thus

(8.3) $$z = \big(f(v, u)u - q(u)v\big)q(v)^{-1}.$$

Then

$$f(u, z) = f\big(u, f(v, u)u - q(u)v\big)q(v)^{-1}$$

(8.4) $$= \big(f(v, u)f(u, u) - q(u)f(u, v)\big)q(v)^{-1}$$

$$= f(u, v)q(u)q(v)^{-1}$$

since $f(u, u) = 2q(u)$. Therefore

$$a\hat{\cdot}v\hat{\cdot}z = avu^{-1}zu^{-1}$$

$$= avu^{\sigma}zu^{\sigma}q(u)^{-2} \qquad\qquad \text{by A1 and 1.3}$$

$$= av\big(u^{\sigma}f(u, z) - z^{\sigma}q(u)\big)q(u)^{-2} \qquad \text{by 1.4 and 3.9}$$

$$= avu^{\sigma}f(u, v)q(u)^{-1}q(v)^{-1}$$

$$\quad - av\big(f(u, v)u^{\sigma} - q(u)v^{\sigma}\big)q(u)^{-1}q(v)^{-1} \quad \text{by 8.3 and 8.4}$$

$$= avv^{\sigma}/q(v) = a \qquad\qquad \text{by A3.}$$

Thus $\hat{\Xi}$ satisfies A3. Notice, too, that

(8.5) $$bv^{\hat{\sigma}}/\hat{q}(v) = buv^{\sigma}uq(v)^{-1}$$

for all $b \in X$ by 3.9 and 8.3.

Choose $b \in X$. Then

$$\hat{h}(a, b\hat{\cdot}v) = h(a, bvu^{-1}u)q(u)^{-1}$$

$$= h(a, bv)q(u)^{-1} \qquad\qquad \text{by A3}$$

$$= h(b, av)q(u)^{-1} + f(h(a, b), 1)vq(u)^{-1} \quad \text{by B2}$$

$$= h(b, avu^{-1}u)q(u)^{-1}$$

$$\quad + f(h(a, bu)q(u)^{-1}, u)vq(u)^{-1} \qquad \text{by A3, B1 and 3.7}$$

$$= \hat{h}(b, a\hat{\cdot}v) + \hat{f}(\hat{h}(a, b), \hat{1})v$$

and

$$\hat{f}(\hat{h}(a\hat{\cdot}v, b), \hat{1}) = f(h(avu^{-1}, bu), u)q(u)^{-2}$$

$$= f(h(avu^{-1}u, bu), 1)q(u)^{-2} \quad \text{by B3}$$

$$= f(h(av, bu), 1)q(u)^{-2} \qquad \text{by A3}$$

$$= f(h(a, bu), v)q(u)^{-2} \qquad \text{by B3}$$

$$= \hat{f}(\hat{h}(a, b), v).$$

Thus $\hat{\Xi}$ satisfies B2 and B3.

Let $\hat{\pi}(a) = \hat{\theta}(a, \hat{1})$ (as in D1). Then

$$\hat{\pi}(a) = q(u)^{-1}\theta(a, u).$$

Hence

$$
\begin{aligned}
a\hat{\cdot}\hat{\theta}(a,v) &= a\theta(a,v)u^{-1}q(u)^{-1} && \text{by A1} \\
&= a\pi(a)vu^{-1}q(u)^{-1} && \text{by D1} \\
&= a\pi(a)uu^{-1}vu^{-1}q(u)^{-1} && \text{by A3} \\
&= a\theta(a,u)u^{-1}vu^{-1}q(u)^{-1} && \text{by D1} \\
&= a\hat{\cdot}\hat{\pi}(a)\hat{\cdot}v && \text{by A1}
\end{aligned}
$$

and $\hat{\pi}(a) \in \langle \hat{1} \rangle$ only if $a = 0$ by 3.11. Thus $\hat{\Xi}$ satisfies D1 and D2. Since

$$
\begin{aligned}
\hat{\theta}(a+b,v) &- \hat{\theta}(a,v) - \hat{\theta}(b,v) \\
&= q(u)^{-1}\big(h(a,bv) - g(a,b)v\big) && \text{by C3} \\
&= q(u)^{-1}\big(h(a,bvu^{-1}u) - g(a,b)v\big) && \text{by A3} \\
&= \hat{h}(a,b\hat{\cdot}v) - q(u)^{-1}g(a,b)v,
\end{aligned}
$$

$\hat{\Xi}$ satisfies C3 with $\hat{g}(a,b) = q(u)^{-1}g(a,b)$.

It remains only to show that $\hat{\Xi}$ satisfies C4. By C1 and 8.2, we have

$$
\begin{aligned}
\hat{\theta}(a,w^{\hat{\sigma}}) &= \theta\big(a, f(w,u)u - q(u)w\big)q(u)^{-2} \\
&= \big(f(w,u)\theta(a,u) - q(u)\theta(a,w)\big)q(u)^{-2}
\end{aligned}
$$

and thus

$$
\begin{aligned}
q(u)^{4}\hat{\theta}(a,w^{\hat{\sigma}})^{\hat{\sigma}}\hat{q}(v) &= q(v)q(u)^{2}\big(f(\hat{\theta}(a,w^{\hat{\sigma}}),u)u - q(u)\hat{\theta}(a,w^{\hat{\sigma}})\big) \\
&= q(v)\bigg(\big(f(w,u)f(\theta(a,u),u) - q(u)f(\theta(a,w),u)\big)u \\
&\qquad\qquad - q(u)\big(f(w,u)\theta(a,u) - q(u)\theta(a,w)\big)\bigg).
\end{aligned}
$$

Next we calculate that

$$
\begin{aligned}
q(u)^{4}\hat{\theta}(a,v)^{\hat{\sigma}}\hat{f}(w,v^{\hat{\sigma}}) &= \big(f(\theta(a,v),u)u - q(u)\theta(a,v)\big) \\
&\qquad \cdot \big(f(w,u)f(v,u) - q(u)f(w,v)\big)
\end{aligned}
$$

and

$$
\begin{aligned}
q(u)^{4}\hat{f}(\hat{\theta}(a,v),w^{\hat{\sigma}})v^{\hat{\sigma}} &= f\big(\theta(a,v), f(w,u)u - q(u)w\big) \\
&\qquad \cdot \big(f(v,u)u - q(u)v\big) \\
&= \big(f(\theta(a,v),u)f(w,u) - f(\theta(a,v),w)q(u)\big) \\
&\qquad \cdot \big(f(v,u)u - q(u)v\big).
\end{aligned}
$$

Hence

$$q(u)^4 \left(\hat{\theta}(a, w^{\hat{\sigma}})^{\hat{\sigma}} \hat{q}(v) - \hat{\theta}(a, v)^{\hat{\sigma}} \hat{f}(w, v^{\hat{\sigma}}) + \hat{f}(\hat{\theta}(a, v), w^{\hat{\sigma}}) v^{\hat{\sigma}} \right)$$

$$= \left(f(\theta(a, u), u) f(w, u) q(v) - f(\theta(a, w), u) q(u) q(v) \right.$$

$$- f(\theta(a, v), u) f(w, u) f(v, u) + f(\theta(a, v), u) f(w, v) q(u)$$

$$\left. + f(\theta(a, v), u) f(w, u) f(v, u) - f(\theta(a, v), w) f(v, u) q(u) \right) u$$

$$+ \left(q(u) f(w, u) f(v, u) - q(u)^2 f(w, v) \right) \theta(a, v)$$

$$- f(w, u) q(u) q(v) \theta(a, u) + q(v) q(u)^2 \theta(a, w)$$

$$+ \left(f(\theta(a, v), w) q(u)^2 - f(\theta(a, v), u) f(w, u) q(u) \right) v.$$

By 3.19.i,iii if $\mathrm{char}(K) = 2$ and by 4.9.i,iii if $\mathrm{char}(K) \neq 2$ (which applies since in this case Ξ is standard), we have

$$f(\theta(a, u), u) f(w, u) = f(\pi(a), 1) q(u) f(w, u)$$

$$= f(\theta(a, u), w) q(u) + f(\theta(a, w), u) q(u).$$

It follows that

$$q(u)^3 \left(\hat{\theta}(a, w^{\hat{\sigma}})^{\hat{\sigma}} \hat{q}(v) - \hat{\theta}(a, v)^{\hat{\sigma}} \hat{f}(w, v^{\hat{\sigma}}) + \hat{f}(\hat{\theta}(a, v), w^{\hat{\sigma}}) v^{\hat{\sigma}} \right)$$

$$= \left(f(\theta(a, u), w) q(v) + f(\theta(a, v), u) f(w, v) \right.$$

$$\left. - f(\theta(a, v), w) f(v, u) \right) u$$

$$+ \left(f(w, u) f(v, u) - q(u) f(w, v) \right) \theta(a, v)$$

$$- f(w, u) q(v) \theta(a, u) + q(v) q(u) \theta(a, w)$$

$$+ \left(f(\theta(a, v), w) q(u) - f(\theta(a, v), u) f(w, u) \right) v.$$

By C2 and several applications of C4, we have

$$q(u)^3 \theta(a \hat{\cdot} v, w) = q(u)^2 \theta(avu^{-1}, w)$$

$$= \theta(avu^{\sigma}, w)$$

$$= \theta(av, w^{\sigma})^{\sigma} q(u) - \theta(av, u^{\sigma})^{\sigma} f(w, u)$$

$$+ f(\theta(av, u^{\sigma})^{\sigma}, w) u + \phi(av, u^{\sigma}) w$$

$$= q(u) \left(\theta(a, w) q(v) - \theta(a, v) f(w, v) + f(\theta(a, v), w) v \right.$$

$$\left. + \phi(a, v) w \right)$$

$$- f(w, u) \left(\theta(a, u) q(v) - \theta(a, v) f(u, v) \right.$$

$$\left. + f(\theta(a, v), u) v + \phi(a, v) u \right)$$

$$+ \left(f(\theta(a, u), w) q(v) - f(\theta(a, v), w) f(u, v) \right.$$

$$\left. + f(\theta(a, v), u) f(v, w) + \phi(a, v) f(u, w) \right) u$$

$$+ \phi(av, u^{\sigma}) w.$$

We conclude, finally, that

$$\hat{\theta}(a,w^{\hat{\sigma}})^{\hat{\sigma}}\hat{q}(v) - \hat{\theta}(a,v)^{\hat{\sigma}}\hat{f}(w,v^{\hat{\sigma}}) + \hat{f}(\hat{\theta}(a,v),w^{\hat{\sigma}})v^{\hat{\sigma}}$$
$$= \hat{\theta}(a\hat{\cdot}v,w) + \Big(\phi(av,u^{\sigma}) + q(u)\phi(a,v)\Big)q(u)^{-3}w.$$

Thus $\hat{\Xi}$ satisfies C4 with $\hat{\phi}(a,v) = \big(\phi(av,u^{\sigma})+q(u)\phi(a,v)\big)q(u)^{-3}$ (since, by 4.5.iii, ϕ is identically zero if $\mathrm{char}(K) \neq 2$).

Suppose that $\mathrm{char}(K) \neq 2$. Then Ξ is standard, so

$$\hat{f}(\hat{\pi}(a),\hat{1}) = q(u)^{-2}f(\theta(a,u),u) = 0$$

for all $a \in X$ by 4.9.i. Therefore $\hat{\Xi}$ is also standard. $\qquad\square$

Definition 8.6. Let $\Xi = (K,L,q,1,X,\cdot,h,\theta)$ be a quadrangular algebra and let $u \in L^*$. If $\mathrm{char}(K) \neq 2$ (in which case Ξ is proper by 3.14), we replace θ by the map obtained by applying 4.2 to Ξ with $\delta = 1/2$ (thus making Ξ standard) and then let Ξ_u denote the quadrangular algebra called $\hat{\Xi}$ in 8.1. If $\mathrm{char}(K) = 2$, we simply let Ξ_u denote the quadrangular algebra called $\hat{\Xi}$ in 8.1.

Definition 8.7. An *isotope* of a quadrangular algebra $\Xi = (K,L,\dots)$ is a quadrangular algebra isomorphic to Ξ_u (as defined in 8.6) for some $u \in L^*$. Two quadrangular algebras are *isotopic* if one is an isotope of the other.

Proposition 8.8. *Let $\Xi = (K,L,q,1,X,\cdot,h,\theta)$ be a quadrangular algebra. Then*

$$(\Xi_u)_v = \Xi_v$$

for all $u,v \in L^*$.

Proof. We can assume that Ξ is standard if $\mathrm{char}(K) \neq 2$. Choose $u,v \in L^*$ and let $\hat{\Xi} = \Xi_u$. Thus

$$\hat{\Xi} = (K,L,\hat{q},\hat{1},X,\hat{\cdot},\hat{h},\hat{\theta})$$

is as in 8.1 and

$$\hat{\Xi}_v = (K,L,\tilde{q},\tilde{1},X,\tilde{\cdot},\tilde{h},\tilde{\theta}),$$

where $\tilde{1} = v$,

$$\tilde{q}(z) = \hat{q}(v)^{-1}\hat{q}(z)$$
$$= q(z)q(v)^{-1};$$
$$a\tilde{\cdot}z = (a\hat{\cdot}z)\hat{\cdot}(v^{\hat{\sigma}}/\hat{q}(v))$$
$$= azu^{-1}(uv^{\sigma}u)q(v)^{-1}u^{-1} \quad \text{by 8.5}$$
$$= azv^{\sigma}q(v)^{-1} = azv^{-1} \quad \text{by A1 and A3;}$$
$$\tilde{h}(a,b) = \hat{h}(a,b\hat{\cdot}v)\hat{q}(v)^{-1}$$
$$= h(a,bvu^{-1}u)q(v)^{-1}$$
$$= h(a,bv)q(v)^{-1} \quad \text{by A1 and A3;}$$

and

$$\tilde{\theta}(a, z) = \hat{q}(v)^{-1}\hat{\theta}(a, z)$$
$$= q(v)^{-1}\theta(a, z)$$

for all $a, b \in X$ and all $z \in L$. Thus $\hat{\Xi}_v = \Xi_v$. $\square$

Proposition 8.9. *Let Ξ be a quadrangular algebra. Then*

(i) *Ξ is an isotope of itself.*
(ii) *If $\hat{\Xi}$ is an isotope of Ξ, then Ξ is an isotope of $\hat{\Xi}$.*
(iii) *If $\tilde{\Xi}$ is an isotope of $\hat{\Xi}$ and $\hat{\Xi}$ is an isotope of Ξ, then $\tilde{\Xi}$ is an isotope of Ξ.*

Proof. Since $\Xi \cong \Xi_1$, this holds by 8.8. $\square$

Chapter Nine

Improper Quadrangular Algebras

In this chapter we show that improper quadrangular algebras are classified by certain indifferent sets (defined in 9.28 below) and certain anisotropic quadratic spaces. First, we have to be a little careful with the definition of "improper."

Proposition 9.1. *Let*

$$\Xi = (K, L, \ldots, \theta)$$

be a quadrangular algebra and let f be as in 1.17. Then f is not identically zero if and only if some isotope of Ξ is proper.

Proof. If $\mathrm{char}(K) \neq 2$, then $f(u, u) = 2q(u) \neq 0$ for all $u \in L^*$ and Ξ is proper by 3.14. We can assume, therefore, that $\mathrm{char}(K) = 2$. Suppose that $f(u, L) \neq 0$ for some $u \in L$. Replacing Ξ by Ξ_u as defined in 8.6, we can assume that $u = 1$. Thus $\sigma \neq 1$ by 1.2. Hence Ξ is proper by 1.27. Conversely, if Ξ_u (for some $u \in L^*$) is proper, then $f(u, L) \neq 0$. $\square$

Definition 9.2. A quadrangular algebra is *improper* if it is not isotopic to a proper quadrangular algebra.

This definition is, we admit, a little odd since it allows some quadrangular algebras to be neither proper nor improper. This problem vanishes, however, if we think in terms of isotopy classes of quadrangular algebras.

Hypothesis 9.3. For the rest of this chapter, we assume that

$$\Xi = (K, L, q, 1, X, \cdot, h, \theta)$$

is an improper quadrangular algebra. Let f, σ, g, ϕ and π be as in 1.17. By 9.1, f is identically zero. By 3.14, we have $\mathrm{char}(K) = 2$. Since Ξ is not proper, we have to be careful not to use results from Chapter 4 (and, in fact, this is why we separated the results in Chapters 3 and 4 into two chapters).

Proposition 9.4. $|K| = \infty$.

Proof. Since f is identically zero and q is anisotropic, q is an injection from L to K. The claim holds, therefore, by 3.13. $\square$

Proposition 9.5. $auv = avu$ *and* $auu = aq(u)$ *for all* $a \in X$ *and all* $u, v \in L$.

Proof. This holds by A3 and 3.8. $\square$

Proposition 9.6. $h(a,b) = h(b,a)$ and $h(au,b) = h(a,bu)$ for all $a,b \in X$ and all $u \in L$.

Proof. Choose $a,b \in X$ and $u \in L$. Then $h(a,bu) = h(b,au)$ by B2. Setting $u = 1$, we obtain $h(a,b) = h(b,a)$. Since a and b are arbitrary, it follows that $h(b,au) = h(au,b)$. $\qquad\square$

Proposition 9.7. *The map g is symmetric.*

Proof. Choose $a,b \in X$. By 9.6, $h(a,b) = h(b,a)$. By C3, therefore, $g(a,b) = g(b,a)$. $\qquad\square$

Proposition 9.8. *Let $a \in X^*$ and $v_1, v_2, \ldots, v_d, w_1, w_2, \ldots, w_e \in L$ (where d and e are arbitrary) and suppose that*

$$av_1 v_2 \cdots v_d = aw_1 w_2 \cdots w_e.$$

Then

$$q(v_1)q(v_2) \cdots q(v_d) = q(w_1)q(w_2) \cdots q(w_e).$$

Proof. By repeated application of C4, we have

$$\pi(a)q(v_1)q(v_2) \cdots q(v_d) \equiv \pi(a)q(w_1)q(w_2) \cdots q(w_e) \pmod{K}.$$

By D2, this implies that

$$q(v_1)q(v_2) \cdots q(v_d) = q(w_1)q(w_2) \cdots q(w_e).$$

$\qquad\square$

Proposition 9.9. $q(\theta(a,u)) = q(\pi(a))q(u)$ *for all $a \in X$ and all $u \in L$.*

Proof. By D1, $a\theta(a,u) = a\pi(a)u$ for all $a \in X$ and all $u \in L$. The claim holds, therefore, by 9.8. $\qquad\square$

Notation 9.10. For each $u \in L$, let ρ_u denote the map from X to itself given by

$$\rho_u(a) = au$$

for all $a \in X$. For each $u \in L^*$, the map ρ_u is (by 3.5) an automorphism of X. Let C denote the subring of $\text{End}(X)$ generated by the automorphisms ρ_u for all $u \in L$. Thus C is the image of the homomorphism from $C(q,1)$ to $\text{End}(X)$ described in 2.22.

Proposition 9.11. $h(aC, aC) = 0$ *for all $a \in X$, where C is as in 9.10.*

Proof. By A1, B1 and 9.6, it suffices to show that

$$h(a, aw_1 w_2 \cdots w_d) = 0$$

for all for all $a \in X^*$, for all $d \geq 1$ and all $w_1, w_2, \ldots, w_d \in L$. We will prove this by induction with respect to d. We consider first the case $d = 1$.

Choose $a \in X^*$ and $u, v, w \in L$. By A1 and C3, we have

$$\theta\big(a(u+v), w\big) = \theta(au + av, w)$$
$$= \theta(au, w) + \theta(av, w) + h(au, avw) + g(au, av)w.$$

By C4, we have

$$\theta\big(a(u+v),w\big) = \theta(a,w)\big(q(u)+q(v)\big) + \phi(a,u+v)w$$

and (by two more applications of C4)

$$\theta(au,w) + \theta(av,w) = \theta(a,w)\big(q(u)+q(v)\big) + \big(\phi(a,u)+\phi(a,v)\big)w.$$

Therefore

$$h(au,avw) = \big(\phi(a,u+v)+\phi(a,u)+\phi(a,v)+g(au,av)\big)w.$$

By 3.19.ii, therefore,

$$h(au,avw) = 0.$$

Setting $u = v = 1$, we conclude (by A2) that $h(a,aw) = 0$.

Now suppose that $d > 1$, choose $w_1, w_2, \ldots, w_d \in L$ and let

$$b = aw_1w_2 \cdots w_{d-1}.$$

By induction, $h(a,b) = 0$. By 3.22 (and 9.4), therefore,

$$b\theta(a,w_d) + b\pi(a)w_d = ah(a,bw_d).$$

By A1 and 9.5, on the other hand,

$$b\theta(a,w_d) + b\pi(a)w_d = \big(a\theta(a,w_d) + a\pi(a)w_d\big)w_1w_2 \cdots w_{d-1}$$

and $a\theta(a,w_d) + a\pi(a)w_d = 0$ by D1. Hence $ah(a,bw_d) = 0$. By A3, therefore,

$$h(a,bw_d) = h(a,aw_1w_2 \cdots w_d) = 0.$$

$\square$

Proposition 9.12. *Let C and ρ_u for $u \in L^*$ be as in 9.10 and let F denote the subring of K generated by $q(L)$. Then F is a subfield of K and there exists a unique isomorphism ψ from C to F which extends the map sending ρ_u to $q(u)$ for each $u \in L^*$.*

Proof. Choose $a \in X^*$, let $d \geq 1$ and let $\xi_1, \ldots, \xi_d$ be d products of finitely many ρ_u's. More precisely, we suppose that for each $j \in [1,d]$,

$$\xi_j = \rho_{u_{1j}}\rho_{u_{2j}} \cdots \rho_{u_{d_j j}}$$

for some $d_j \geq 1$ and some $u_{1j}, u_{2j}, \ldots, u_{d_j j} \in L^*$. By C3 and 9.11,

$$(9.13) \qquad \pi\left(\sum_{j=1}^{d}\xi_j(a)\right) \equiv \sum_{j=1}^{d}\pi(\xi_j(a)) \pmod{K}$$

for all $i,j \in [1,d]$. By C4,

$$(9.14) \qquad \sum_{j=1}^{d}\pi(\xi_j(a)) \equiv \pi(a)\left(\sum_{j=1}^{d}q(u_{1j})q(u_{2j}) \cdots q(u_{d_j j})\right) \pmod{K}$$

for all $j \in [1,d]$. By D2, 9.13 and 9.14, it follows (since a is arbitrary) that

$$\xi_1 + \xi_2 + \cdots + \xi_d = 0$$

if and only if

$$\sum_{j=1}^{d} q(u_{1j})q(u_{2j}) \cdots q(u_{d_jj}) = 0.$$

This implies that there is a unique isomorphism ψ (of rings) from C to F sending ρ_u to $q(u)$ for all $u \in L^*$.

Let $t \in F^*$. Then t is a sum of products of $q(u)$'s and $t^{-2}q(u) = q(t^{-1}u)$ for each $u \in L$. Thus $t^{-2}t$ is also a sum of products of $q(u)$'s. In other words, $t^{-1} \in F$. Thus the subring F of K is, in fact, a subfield. $\qquad\square$

Corollary 9.15. *Let C be in 9.10. Then $t \mapsto \rho_t$ is an isomorphism from K to a subfield of C and X is a vector space over C.*

Proof. If $t \in K$, then $\rho_t(a) = at$ by A1 and A2. Thus $t \mapsto \rho_t$ is an isomorphism from K into C. By 9.12, C is a field. By 9.10, X is a vector space over C. $\qquad\square$

Corollary 9.16. *Let $a, b \in X$ and let C be as in 9.10. Then either $aC = bC$ or $aC \cap bC = 0$.*

Proof. This holds by 9.15. $\qquad\square$

Proposition 9.17. *Suppose that h is not identically zero. Then*

$$F = q(L),$$

where F is as in 9.12; there is a unique multiplication on L (which we will denote by juxtaposition) such that

(i) *L (endowed with this multiplication) is a field containing K as a subfield and*

(ii) *q is an isomorphism from L to F;*

and X is a vector space over L (endowed with this multiplication) with respect to the map $(a, u) \mapsto au$.

Proof. Choose $a, b \in X$ such that $h(a, b) \neq 0$. By 9.11,

$$h(a, aC) = 0.$$

Thus $b \notin aC$. By 9.16, therefore, $aC \cap bC = 0$. By 3.22 (and 9.4), it follows that

(9.18) $ah(a, b)u = ah(a, bu)$

for all $u \in L$. Setting $u = h(a, b)^{-1}$ in 9.18 and applying A3, we have

$$a = ah(a, bh(a, b)^{-1}).$$

Now let $w_1, w_2, \ldots, w_d \in L$. By repeated application of 9.18, it follows that

$$aw_1w_2 \cdots w_d = ah(a, bh(a, b)^{-1})w_1w_2 \cdots w_d = av$$

for

$$v = h\big(a, bh(a, b)^{-1}w_1w_2 \cdots w_d\big).$$

By 9.8, therefore,

$$q(w_1)q(w_2)\cdots q(w_d) = q(v) \in q(L).$$

Since every element of F is a sum of such products (and since q is additive and thus $q(L)$ is closed under addition), it follows that $F = q(L)$. Thus by A1, 3.4 and 9.12, the map $u \mapsto \rho_u$ is an additive bijection from L to

$$C = \{\rho_u \mid u \in L\}.$$

There is a unique multiplication on L which makes this map multiplicative. By 9.15, this multiplication extends the multiplication on $K \subset L$ and X is a vector space over L (endowed with this multiplication) with respect to the map $(a, u) \mapsto au$.   □

Lemma 9.19. *Let $a, b, c \in X$. Then*

   (i) $g(a, b) + g(a + b, c) = g(a, b + c) + g(b, c)$;
   (ii) $g(a, a) = 0$; and
  (iii) $g(a, 0) = g(0, a) = 0$.

Proof. Expanding both sides of the identity

$$\pi\big((a + b) + c\big) = \pi\big(a + (b + c)\big)$$

by C3 and applying B1, we obtain (i). By 9.11, $h(a, a) = 0$. By 3.15, therefore, (ii) holds. By B1, we have $h(a, 0) = h(0, a) = 0$. By C2, we have $\pi(0) = 0$. By C3, therefore, (iii) holds.   □

Lemma 9.20. *Suppose that h is not identically zero and let L be endowed with the multiplication described in 9.17. Then*

   (i) $\phi(a, 1) = 0$;
  (ii) $\phi(a, uv) = \phi(a, u)q(v) + \phi(au, v)$; and
 (iii) $\theta(a, v) = \pi(a)v$ and $h(a, bv) = h(a, b)v$

for all $a, b \in X$ and all $u, v \in L$.

Proof. Choose $a, b \in X$ and $u, v \in L$ and let $w = uv$. By A2, we have $\pi(a \cdot 1) = \pi(a)$. By C4, therefore, (i) holds. By 9.17, $auv = aw$ and $q(u)q(v) = q(w)$. By three applications of C4, we thus have

$$\begin{aligned}
\pi(a)q(w) + \phi(a, w) &= \pi(aw) \\
&= \pi(auv) \\
&= \pi(au)q(v) + \phi(au, v) \\
&= \pi(a)q(u)q(v) + \phi(a, u)q(v) + \phi(au, v) \\
&= \pi(a)q(w) + \phi(a, u)q(v) + \phi(au, v).
\end{aligned}$$

Therefore (ii) holds.

To prove (iii), it suffices to assume (by B1 and 3.12) that $a \neq 0$. By D1, $a\theta(a, v) = a(\pi(a)v)$. Therefore $\theta(a, v) = \pi(a)v$ by 3.4. Thus $b\theta(a, v) = b\pi(a)v$. By 3.22 (and 9.4), it follows that

$$ah(a, bv) = a(h(a, b)v).$$

Hence $h(a, bv) = h(a, b)v$, again by 3.4. Thus (iii) holds.   □

Proposition 9.21. *Suppose that h is not identically zero and let L be endowed with the multiplication described in 9.17. Let $V = X \times K$ (as a set only) and let*

$$(a,t) + (b,s) = \big(a + b, s + t + g(a,b)\big)$$

for all $(a,t),(b,s) \in V$. Then V endowed with this addition is a vector space over L with scalar multiplication given by

$$(a,t)u = \big(au, tq(u) + \phi(a,u)\big)$$

for all $(a,t) \in V$ and all $u \in L$.

Proof. By 9.7 and 9.19, $(V, +)$ is an abelian group with identity $(0,0)$ and $-(a,t) = (a,t)$ for all $(a,t) \in V$. By 9.20.i, $(a,t)1 = (a,t)$ for all $(a,t) \in V$. Choose $(a,t),(b,s) \in V$ and $u,v \in L^*$. Then

$$\big((a,t)u\big)v = (a,t)(uv)$$

by 9.17 and 9.20.ii,

$$(a,t)(u + v) = (a,t)u + (a,t)v$$

by A1 and 3.19.ii and

$$\big((a,t) + (b,s)\big)u = (a,t)u + (b,s)u$$

by A1 and 3.20. $\qquad\qquad\qquad\qquad\qquad\qquad\qquad\qquad\qquad\qquad\quad\square$

Proposition 9.22. *Suppose that h is not identically zero. Then there exists a map τ from X to K such that*

(i) $\tau(a + b) = \tau(a) + \tau(b) + g(a,b)$;
(ii) $\tau(au) = \tau(a)q(u) + \phi(a,u)$; and
(iii) $\tau(at) = \tau(a)t^2$

for all $a,b \in X$, all $u \in L$ and all $t \in K$.

Proof. Let V be as in 9.21 and let W denote the subspace

$$\{(0,t) \mid t \in K\}.$$

Since V is a vector space over L, there exists a subspace U of V such that $V = U + W$ and $U \cap W = 0$ (i.e. U is a *complement* of W in V). Suppose that (a,s) and (a,t) are two elements of U with the same first coordinate. Then

$$(a,s) + (a,t) = (0, s + t)$$

by 9.19.ii and hence $s = t$ since $U \cap W = 0$. There thus exists a map τ from X to K such that

$$U = \{(a, \tau(a)) \mid a \in X\}.$$

Now choose $a,b \in X$, $u \in L$ and $t \in K$. We have

$$(a, \tau(a)) + (b, \tau(b)) = (a + b, \tau(a + b))$$

since U is closed under addition. Therefore (i) holds. We have

$$(a, \tau(a))u = (au, \tau(au))$$

because U is closed under scalar multiplication. Therefore (ii) holds. By C1, C2 and C4, we have $\phi(a, ut) = \phi(a, u)t^2$. Setting $u = t \cdot 1$ in (ii), we have (by A1 and A2)

$$\tau(at) = \tau(a)t^2 + \phi(a, 1)t^2.$$

By 9.20.i, therefore, (iii) holds.   □

Notation 9.23. Suppose that h is not identically zero and let τ be as in 9.22. Let

$$\tilde{\Xi} = (\hat{K}, \hat{L}, \ldots, \hat{\theta})$$

be the equivalent quadrangular algebra obtained from Ξ by setting $p = \tau$ and $\omega = 1$ in 1.24 and let $\hat{g}$ and $\hat{\phi}$ be as in C3 and C4. By 1.24, 9.22.i and 9.22.ii, both $\hat{g}$ and $\hat{\phi}$ are identically zero.

Proposition 9.24. *Suppose that h is not identically zero, let L be endowed with the multiplication described in 9.17, replace Ξ by the equivalent quad-rangular algebra $\hat{\Xi}$ described in 9.23 (so g and ϕ are both identically zero) and let V be as in 9.21 (so V is now the direct sum of X and K). Let Q be the map from V to L given by*

$$Q(a, t) = \pi(a) + t$$

for all $(a, t) \in V$. Then Q is an anisotropic (but defective) quadratic form on V (as a vector space over L as described in 9.21) whose associated bilinear form H is given by

$$H\big((a, t), (b, s)\big) = h(a, b)$$

for all $(a, t), (b, s) \in V$.

Proof. Let $(a, t), (b, s) \in V$ and let $u \in L$. By B1, 9.6 and 9.20.iii, H is a bilinear form on V. We have

$$\begin{aligned}
Q\big((a, t) + (b, s)\big) &= Q((a + b, s + t) \\
&= \pi(a + b) + s + t \\
&= Q(a, t) + Q(b, s) + H\big((a, t), (b, s)\big)
\end{aligned}$$

by C3 since g is identically zero and

$$\begin{aligned}
Q(a, t)u^2 &= (\pi(a) + t)q(u) & \text{by 9.5} \\
&= \pi(a)q(u) + tq(u) \\
&= \pi(au) + tq(u) \\
&= Q\big(au, tq(u)\big) = Q(((a, t)u)
\end{aligned}$$

by C4 since ϕ is identically zero. Thus Q is a quadratic form on V. By D2, Q is anisotropic. Since (by B1) $H\big((a, t), (0, s)\big) = 0$ for all $(a, t) \in V$ and all $s \in K$, Q is defective.   □

To state the converse of 9.24, we suspend 9.3 for a moment:

Proposition 9.25. *Let*

$$(L, X, \pi)$$

be a quadratic space such that char$(L) = 2$, *let* h *denote the corresponding bilinear form, let*

$$(a, u) \mapsto a \cdot u$$

denote the scalar multiplication from $X \times L$ *to* L, *let* K *be a subfield of* L *such that*

$$L^2 \subset K \subset L,$$

let q *be the quadratic form on* L *as a vector space over* K *given by* $q(u) = u^2$ *for all* $u \in L$, *let* X *be considered as a vector space over* K *by restriction of scalars, let* $\theta(a, u) = \pi(a)u$ *for all* $a \in X$ *and all* $u \in L$ *and let*

$$\Xi = (K, L, q, 1, X, \cdot, h, \theta).$$

Suppose that

$$(L, X, \pi) \perp (L, K, q_K)$$

is anisotropic, where (L, K, q_K) *is as in 2.14 with* L *in place of* K *and* K *in place of* F. *Then* Ξ *is an improper quadrangular algebra.*

Proof. The hypothesis that

$$(L, X, \pi) \perp (L, K, q_K)$$

is anisotropic is equivalent to saying that $\pi(a) \in K$ if and only if $a = 0$. Thus Ξ satisfies D2. It is clear that Ξ satisfies all the other axioms of a quadrangular algebra. $\square$

Theorem 9.26. *Improper quadrangular algebras*

$$\Xi = (K, L, \dots, h, \theta)$$

such that h *is not identically zero are classified by pairs*

$$\big(K, (L, X, \pi)_K\big),$$

where (L, X, π) *is a quadratic space whose corresponding bilinear form is not identically zero;* K *is a subfield of* L *such that*

(i) $L^2 \subset K$ *and*
(ii) $(L, X, \pi) \perp (L, K, q_K)$ *is anisotropic, where* (L, K, q_K) *is as in 2.14;*

and $(L, X, \pi)_K$ *denotes the set of quadratic spaces*

$$\{(K, L, \alpha\pi) \mid \alpha \in K^*\}.$$

Proof. If we start with Ξ and the assumption that h is not identically zero and endow L with the multiplication described in 9.17, then by 9.24, (L, X, π) is a quadratic space fulfilling the hypothesis of 9.25. If we apply 9.25 to K and (L, X, π) (or to K and any quadratic space in $(L, X, \pi)_K$), we recover Ξ (up to isomorphism). Suppose, instead, that we start with a quadratic space (L, X, π) and a subfield K of L satisfying the hypotheses of 9.25 such that the bilinear form associated with π is not identically zero. Let Ξ be the quadrangular algebra obtained from applying 9.25 to this data. If we apply 9.24 to Ξ or to a quadrangular algebra isomorphic to it, we recover K and a quadratic space in $(L, X, \pi)_K$. □

Remark 9.27. Let

$$(L, X_0, \pi_0)$$

be an arbitrary anisotropic but defective quadratic space. Let h_0 denote the bilinear form associated with π_0, choose a non-zero element u in the defect of π_0 (i.e. in the radical of h_0), choose a subspace X of X_0 such that

$$X_0 = X \oplus \langle u \rangle,$$

let π denote the restriction of $\pi_0/\pi_0(u)$ to X, let $K = L^2$ and let π_1 denote the restriction of $\pi_0/\pi_0(u)$ to $\langle u \rangle$. Then $tu \mapsto t^2$ is an L-linear isomorphism from $(L, \langle u \rangle, \pi_1)$ to (L, K, q_K) and hence

$$\begin{aligned}
\left(L, X_0, \pi_0/\pi_0(u)\right) &\cong (L, X, \pi) \perp (L, \langle u \rangle, \pi_1) \\
&\cong (L, X, \pi) \perp (L, K, q_K)
\end{aligned}$$

since $\langle u \rangle^\perp = X_0$. Thus the pair (L, X, π) and K satisfy the hypotheses of 9.25 as long as h_0 is not identically zero.

We return now to the assumption 9.3. For the rest of this chapter, we examine the case that h is identically zero. The following definition is taken from (10.1) and (10.7) in [12].

Definition 9.28. An *indifferent set* is a triple (F, L_0, X_0) such that F is a field in characteristic two and L_0 and X_0 are additive subgroups of F both containing 1 such

(i) $L_0^2 X_0 \subset X_0$ and $X_0 L_0 \subset L_0$ and
(ii) L_0 generates F as a ring.

Proposition 9.29. *Let (F, L_0, X_0) be an indifferent set. Then*

(i) $F^2 \subset X_0 \subset L_0$,
(ii) $F^2 L_0 \subset L_0$ and $F^2 X_0 \subset X_0$.

Proof. Since $L_0^2 X_0 \subset X_0$ and every element of F is a sum of products of elements of L_0, we have $F^2 X_0 \subset X_0$. Since $1 \in X_0$, we therefore have

$$F^2 \subset F^2 X_0 \subset X_0$$

and thus

$$F^2 L_0 \subset X_0 L_0 \subset L_0.$$

Thus (ii) holds. Since $1 \in L_0$, we have $X_0 \subset X_0 L_0 \subset L_0$. Thus (i) holds. □

Proposition 9.30. *Suppose that h is identically zero, let $L_0 = q(L)$, let*

$$X_0 = \{q(\pi(a)) + t^2 \mid a \in X,\ t \in K\}$$

(so $K^2 \subset X_0$) and let F denote the subring of K generated by L_0. Then (F, L_0, X_0) is an indifferent set such that $X_0 \neq K^2$.

Proof. Since f is identically zero, L_0 is an additive subgroup of K. Since h is identically zero, X_0 is also (by C3) an additive subgroup of K. Since $1 = q(1) = q(\pi(0)) + 1^2$, both L_0 and X_0 contain 1. Let $a \in X$, $t \in K$ and $u \in L$. Then

$$
\begin{aligned}
q(u)^2 \big(q(\pi(a)) + t^2 \big) &= q\big(\pi(a)q(u) + q(u)t\big) \\
&= q\big(\pi(au) + \phi(a, u) + q(u)t\big) \\
&= q(\pi(au)) + \big(\phi(a, u) + q(u)t\big)^2 \in X_0
\end{aligned}
$$

by C4 and

$$
\begin{aligned}
\big(q(\pi(a)) + t^2\big)q(u) &= q(\pi(a))q(u) + q(tu) \\
&= q(\theta(a, u) + tu) \in L_0
\end{aligned}
$$

by 9.9. Thus (F, L_0, X_0) satisfies 9.28.i. By 9.12, (F, L_0, X_0) also satisfies 9.28.ii. Suppose, finally, that $q(\pi(a)) = t^2$. Then $q(\pi(a) + t) = 0$ and thus $a = 0$ by D2. Since X is non-trivial, we conclude that $X_0 \neq K^2$. $\qquad\square$

We now suspend 9.3 for the rest of the chapter.

Proposition 9.31. *Let (F, L_0, X_0) be an indifferent set and let K be a field such that*

$$K^2 \subset X_0 \subset F \subset K$$

but $X_0 \neq K^2$. Thus

(i) *$K^2 L_0 \subset X_0 L_0 \subset L_0$ (by 9.28.i), so L_0 is a vector space over K with respect to the scalar multiplication $*$ (from $L_0 \times K$ to L_0) given by*

$$u * t = ut^2$$

for all $u \in L_0$ and all $t \in K$;

(ii) *$F^2 X_0 \subset X_0$ (by 9.29.ii), so X_0 is a vector space over F with respect to the scalar multiplication $\bullet$ obtained by restricting $*$ to $X_0 \times F \subset L_0 \times K$;*

(iii) *K^2 is a subspace of X_0 with respect to $\bullet$.*

Let X be a complement of K^2 in X_0 (as a vector space over F with respect to $\bullet$) and let $\circ$ denote the restriction of $\bullet$ to $X \times L_0 \subset X_0 \times F$. Then

(iv) *X is also a vector space over K with respect to the scalar multiplication $\diamond$ given by*

$$a \diamond t = a \bullet t^2 = at^4$$

for all $a \in X$ and all $t \in K$.

Let $L = L_0$; let q be the identity map from L to itself; let h be the map from $X \times X$ to L which maps everything to zero; let $\theta(a, u) = au$ for all $a \in X$ and all $u \in L$; and let

$$\Xi = (K, L, q, 1, X, \circ, h, \theta).$$

Then Ξ is a quadrangular algebra which does not depend (up to equivalence) on the choice of the complement X.

Proof. Choose $a, b \in X$, $u, w \in L^*$ and $t \in K$. Since

$$q(u * t) = ut^2,$$

q is a quadratic form. Since

$$(a \diamond t) \circ u = at^4 u^2 = (a \circ u) \diamond t = a \circ (u * t)$$

and $a \circ 1 = a$, Ξ satisfies A1 and A2. Since

$$(a \circ u) \circ u^{-1} = (au^2) \circ (u * q(u)^{-1})$$
$$= (au^2) \circ (u^{-1}) = au^2 u^{-2} = a,$$

Ξ satisfies A3. Since

$$\theta(a, u * t) = aut^2 = au * t = \theta(a, u) * t$$

and

$$\theta(a \diamond t, u) = at^4 u = (au) * t^2 = \theta(a, u) * t^2,$$

Ξ satisfies C1 and C2. Since

$$\theta(a + b, u) = (a + b)u = au + bu = \theta(a, u) + \theta(b, u)$$

and

$$\theta(a \circ u, w) = au^2 w = aw * q(u) = \theta(a, w) * q(u),$$

Ξ satisfies C3 (with g identically zero) and Ξ satisfies C4 (with ϕ identically zero). Since

$$a \circ \theta(a, u) = a \circ (au) = \big(a \circ \pi(a)\big) \circ u,$$

Ξ satisfies D1. If $\pi(a) \in 1 * K = K^2$, then $a \in X \cap K^2 = 0$. Thus Ξ satisfies D2. We conclude that Ξ is a quadrangular algebra.

Suppose that $\hat{X}$ is another complement of K^2 in X_0 (as a vector space over F with respect to $\bullet$) and let $\hat{\Xi}$ denote the quadrangular algebra we obtain with this choice. There exists a linear automorphism ψ of X_0 which maps X to $\hat{X}$ such that

$$\psi(a) \equiv a \pmod{K^2}$$

for all $a \in X$. Since ψ is linear with respect to $\bullet$, we have

$$\psi(a \circ u) = \psi(a) \circ u$$

for all $a \in X$ and all $u \in L$. We also have

$$\theta(a, u) = au \equiv \psi(a)u \pmod{u * K}$$
$$\equiv \theta(\psi(a), u) \pmod{u * K}$$

for all $a \in X$ and all $u \in L$. Thus the identity maps on K and L combine with ψ to yield an isomorphism (as defined in 1.25 with $\hat{\omega} = 1$) from Ξ to $\hat{\Xi}$. $\square$

Proposition 9.32. *Let* (F, L_0, X_0) *be an indifferent set and let* K *be a field such that*

$$K^2 \subset X_0 \subset F \subset K$$

and $X_0 \neq K^2$. *Then for each* $\omega \in K^*$, $(F, L_0, \omega^2 X_0)$ *is an indifferent set,* $K^2 \subset \omega^2 X_0$ *and* $\omega^2 X_0 \neq K^2$.

Proof. Let $\omega \in K^*$. By 9.29.ii, $\omega^{-2} = \omega^{-4}\omega^2 \in F^2 X_0 \subset X_0$. Thus $1 \in \omega^2 X_0$. By 9.28.i, we have $\omega^2 X_0 L_0 \subset \omega^2 L_0 \subset X_0 L_0 \subset L_0$. Hence $(F, L_0, \omega^2 X_0)$ is an indifferent set. Since $K^2 = \omega^2 K^2$, we have $K^2 \subset \omega^2 X_0$ and $\omega^2 X_0 \neq K^2$. $\qquad\square$

Theorem 9.33. *Improper quadrangular algebras*

$$\Xi = (K, L, \dots, h, \theta)$$

such that h *is identically zero are classified by pairs*

$$(K, (F, L_0, X_0)_K),$$

where (F, L_0, X_0) *is an indifferent set,* K *is a field such that*

$$K^2 \subset X_0 \subset F \subset K,$$

$X_0 \neq K^2$ *and* $(F, L_0, X_0)_K = \{(F, L_0, \omega^2 X_0) \mid \omega \in K^*\}$; *see 9.32.*

Proof. If we start with $\Xi = (K, \dots)$ and the assumption that h is identically zero and let (F, L_0, X_0) be the indifferent set obtained by applying 9.30 to Ξ, then Ξ can be recovered (up to isomorphism) by applying 9.31 to K and (F, L_0, X_0) (or to K and any indifferent set in $(K, L_0, X_0)_K$). Conversely, if we start with a pair

$$(K, (F, L_0, X_0)),$$

where (F, L_0, X_0) is an indifferent set and K is a field such that

$$K^2 \subset X_0 \subset F \subset K$$

and $X_0 \neq K^2$ and let Ξ be the quadrangular algebra (or one isomorphic to it) obtained by applying 9.31 to this data, then we recover K and an indifferent set in $(F, L_0, X_0)_K$ by applying 9.30 to Ξ. $\qquad\square$

Chapter Ten

Existence

In this chapter, we give a proof of the following result:

Theorem 10.1. *Let (K, L, q) be a quadratic space of type E_6, E_7, E_8 or F_4 and let $u \in L^*$. Then there exist X, $\cdot$, h and θ such that*

$$\Xi = (K, L, q/q(u), u, X, \cdot, h, \theta)$$

is a quadrangular algebra.

By 8.1, it suffices to show in 10.1 that Ξ exists for *some* $u \in L^*$.

We have already shown in 3.2 and 3.3 that the quadrangular algebra Ξ in 10.1 is unique.

The proof of 10.1 we give is based on results in Chapters 13 and 14 of [12]. Our strategy is to construct a right $C(q/q(u), u)$-module X explicitly (using coordinates), give formulas for the maps h and θ in terms of these coordinates and then check that the axioms of a quadrangular algebra hold by calculation. In the case that (K, L, q) is of type E_6, E_7 or E_8, this requires extensive calculations which have, in fact, already been carried out in [12]. Where it seems appropriate (mainly in the proof of 10.16 below), we will simply indicate how to interpret the relevant parts of [12] in the language of quadrangular algebras rather than reproduce the details here.

We begin the proof of 10.1. As usual, we let f be as in 1.1.i. Suppose first that (K, L, q) is a quadratic space of type E_ℓ for $\ell = 6$, 7 or 8. Let

$$d = \begin{cases} 3 & \text{if } \ell = 6, \\ 4 & \text{if } \ell = 7, \\ 6 & \text{if } \ell = 8. \end{cases}$$

By 2.13, (K, L, q) is anisotropic and we can identify L with a vector space over a field E having a basis $v_1, v_2, \ldots, v_d$ such that

(i) E/K is a separable quadratic extension with norm N;

(ii) $q(u_1 v_1 + u_2 v_2 + \cdots + u_d v_d) = \alpha_1 N(u_1) + \alpha_2 N(u_2) + \cdots + \alpha_d N(u_d)$

for all $u_1, u_2, \ldots, u_d \in E$, where

(10.2)
$$\alpha_i = q(v_i)$$

for all $i \in [1, d]$;

(iii) $\alpha_1 \alpha_2 \alpha_3 \alpha_4 \notin N(E)$ if $\ell = 7$; and

(iv) $-\alpha_1 \alpha_2 \alpha_3 \cdots \alpha_6 \in N(E)$ if $\ell = 8$.

We set $1 = v_1$ and replace q by $q/q(v_1)$. Thus $\alpha_1 = 1$ from now on. We have $f(1, uv_i) = 0$ for all $u \in E$ and all $i \in [2, d]$. By 1.2 and 1.12, therefore,

$$(10.3) \qquad (uv_i)^\sigma = \begin{cases} u^\tau v_i & \text{if } i = 1, \\ -uv_i & \text{if } i > 1 \end{cases}$$

for all elements $u \in E$, where τ denotes the unique non-trivial element of $\mathrm{Gal}(E/K)$.

By 2.21, we have the following identities in $C(q, 1)$, where we denote multiplication by juxtaposition:

$$(10.4) \qquad uu = u(f(1, u) - u^\sigma) = uf(u, 1) - q(u)$$

and

$$(10.5) \qquad \begin{aligned} f(u, v) &= q(u + v) - q(u) - q(v) \\ &= (u + v)(u + v)^\sigma - uu^\sigma - vv^\sigma \\ &= uv^\sigma + vu^\sigma \end{aligned}$$

for all $u, v \in L$. Thus, in particular,

$$(10.6) \qquad uv \equiv -vu \pmod{\langle 1, u, v \rangle}$$

in $C(q, 1)$ for all $u, v \in L$ by 1.2 and 10.5.

By 2.24, $\dim_K C(q, 1) = 2^{2d-1}$. Let $w_1, w_2, \ldots, w_{2d}$ be an ordered basis of L over K such that $w_i = v_i$ for all $i \in [1, d]$. Let B_0 denote the subset of $C(q, 1)$ containing $1 = v_1$ and all products of distinct elements in $w_2, \ldots, w_{2d}$ multiplied in ascending order. By 10.4 and 10.6, every product of elements of L in $C(q, 1)$ can be written as a linear combination over K of the elements of B_0. Thus $C(q, 1)$ is spanned by B_0. Since $|B_0| = 2^{2d-1}$, B_0 is, in fact, a basis of $C(q, 1)$.

Notation 10.7. Let I denote the set of all subsets of the interval $[2, d]$. We write each element of $I \setminus \{\emptyset\}$ as a sequence

$$i_1 i_2 \cdots i_m$$

in ascending order. For each $x \in I$, let

$$v_x = \begin{cases} v_1 & \text{if } x = \emptyset, \\ v_{i_1} v_{i_2} \cdots v_{i_m} & \text{if } x = i_1 i_2 \cdots i_m. \end{cases}$$

Let

$$\tilde{B} = \{v_x \mid x \in I\}$$

and let $|v_x|$ denote the cardinality of x for all $x \in I$ (so $|v_1| = 0$). We also set $\alpha_\emptyset = 1$ and

$$\alpha_x = \alpha_{i_1} \alpha_{i_2} \cdots \alpha_{i_m}$$

for every non-empty element $x = i_1 i_2 \cdots i_m \in I$, where $\alpha_2, \ldots, \alpha_d$ are as in 10.2. We have $|\tilde{B}| = 2^{d-1}$ and the set $\tilde{B}$ is linearly independent over K (since B_0 is a basis of $C(q, 1)$).

By 10.4 and 10.5,

$$(10.8) \qquad\qquad v_i v_i = -\alpha_i$$

for all $i \in [2, d]$ and

$$(10.9) \qquad\qquad v_i v_j = -v_j v_i$$

whenever $i, j \in [2, d]$ are distinct. By 10.8 and 10.9, the product of any two elements in $\tilde{B}$ is an element of $\tilde{B}$ multiplied by a scalar of the form ± 1 times a product of some of the α_i. We set $|tb| = |b|$ for all $b \in \tilde{B}$ and all $t \in K^*$. Thus

$$|bv_i| = |b| \pm 1$$

for all $b \in \tilde{B}$ and all $i \in [2, d]$.

Now let $\tilde{X}$ be a vector space over E with basis $\tilde{B}$. Thus

$$\dim_K \tilde{X} = 2|\tilde{B}| = 2^d.$$

We think of $\tilde{B}$ simply as a set here, not as a subset of $C(q, 1)$, so $\tilde{X}$ is not a subset of $C(q, 1)$ and we do not have a multiplication on $\tilde{X}$. We do define products of the form bv_i in $\tilde{X}$ for $b \in \tilde{B}$ and $i \in [2, v]$, however, as the scalar multiple (by a scalar in K) of an element of $\tilde{B}$ obtained by multiplying b with v_i in $C(q, 1)$. A product of the form $r \cdot bv_i$ (or just rbv_i) for $r \in E$, $b \in \tilde{B}$ and $i \in [2, d]$ is to be interpreted as the scalar multiple in $\tilde{X}$ by r of the element bv_i.

Definition 10.10. Let $(a, v) \mapsto a \cdot v$ be the unique map from $\tilde{X} \times L$ to $\tilde{X}$ which is bilinear over K such that

$$ub \cdot rv_i = \begin{cases} ur^{\tau^{|b|}} \cdot b & \text{if } i = 1, \\ ur \cdot bv_i & \text{if } i \in [2, d] \text{ and } |bv_i| > |b|, \\ ur^\tau \cdot bv_i & \text{if } i \in [2, d] \text{ and } |bv_i| < |b| \end{cases}$$

for all $u, r \in E$, all $b \in \tilde{B}$ and all $i \in [1, d]$, where the product bv_i is defined as above and τ is as in 10.3.

Proposition 10.11. $(K, L, q, 1, \tilde{X}, \cdot)$ *satisfies axioms A1, A2 and A3.*

Proof. Axioms A1 and A2 hold by definition. Choose $u \in E$ and $b \in \tilde{B}$. To prove A3, it suffices to show that

$$(ub \cdot v) \cdot v^\sigma = uq(v)b$$

for all $v \in L$. Let $r, z \in E$. By 10.3, 10.8 and 10.10, we have

$$(ub \cdot rv_1) \cdot (rv_1)^\sigma = (ub \cdot rv_1) \cdot (r^\tau v_1)$$
$$= urr^\tau b = ubN(r)$$

and

$$(ub \cdot rv_i) \cdot (rv_i)^\sigma = (-ub \cdot rv_i) \cdot (rv_i)$$
$$= u\alpha_i rr^\tau b = u\alpha_i N(r)b$$

for all $i \in [2, d]$. Also

$$(ub \cdot rv_1) \cdot zv_i = (ub \cdot zv_i) \cdot r^\tau v_1$$

for all $i \in [2, d]$ (since $|bv_i| = |b| \pm 1$) and

$$(ub \cdot rv_i) \cdot zv_j = (-ub \cdot zv_j) \cdot rv_i$$

for all distinct $i, j \in [2, d]$ (since $bv_i v_j = -bv_j v_i$ by 10.9). It follows that

$$
\begin{aligned}
ub \cdot (u_1 v_1 + \cdots + u_d v_d) \cdot (u_1 v_1 + \cdots + u_d v_d)^\sigma \\
= ub \cdot (u_1 v_1 + \cdots + u_d v_d) \cdot (u_1^\tau v_1 - u_2 v_d - \cdots - u_d v_d) \\
= ub\big(N(u_1) + \alpha_2 N(u_2) + \cdots + \alpha_d N(u_d)\big) \\
= ubq(u_1 v_1 + \cdots + u_d v_d)
\end{aligned}
$$

for all $u_1, \ldots, u_d \in E$.   □

Now suppose that $\ell = 8$. Then $-\alpha_{23456} = N(u)$ for some $u \in E^*$. Replacing v_6 by $u^{-1} v_6$, we can assume that $\alpha_{23456} = -1$. Let sgn denote the unique map from I to $\{+1, -1\}$ such that

$$v_{23456} v_x = \text{sgn}(x)\alpha_x v_{x'}$$

in $C(q, 1)$ for all $x \in I$, where x' denotes the complement of x in $[2, d]$. Let

$$I_2 = \{x \in I \mid |x| \le 2\}$$

and let

$$T = \{e v_x - \text{sgn}(x)\alpha_x e^\tau v_{x'} \mid e \in E \text{ and } x \in I_2\}.$$

By 10.10 and some calculation, we verify that $T \cdot rv_i \subset T$ for all $r \in E$ and all $i \in [1, d]$. Let M be the K-subspace (not the E-subspace!) of $\tilde{X}$ spanned by the set T. Thus $M \cdot L \subset M$, so the map $(a, v) \mapsto a \cdot v$ defined in 10.10 induces a map we also denote by $(a, v) \mapsto a \cdot v$ from $\tilde{X}/M \times L$ to $\tilde{X}/M$. We have $\dim_K M = 2|I_2| = 32$ and thus

$$\dim_K \tilde{X}/M = 32.$$

Let

(10.12)
$$X = \begin{cases} \tilde{X} & \text{if } \ell = 6 \text{ or } 7, \\ \tilde{X}/M & \text{if } \ell = 8 \end{cases}$$

and let $B = \tilde{B}$ if $\ell = 6$ or 7. If $\ell = 8$, we let B denote the image of

$$\{b \in \tilde{B} \mid |b| \le 2\}$$

in X and let ub for $u \in E$ and $b \in \tilde{B}$ denote the image of ub in X. Thus every element of X can be written uniquely as a sum of elements of the form ub for $u \in E$ and $b \in B$ (for all three values of ℓ). By 10.11,

$$(K, L, q, 1, X, \cdot)$$

satisfies the axioms A1, A2 and A3 and $\dim_K X$ is as in 2.27.

Choose γ in E such that

$$\gamma + \gamma^\sigma = \begin{cases} 0 & \text{if char}(K) \neq 2, \\ 1 & \text{if char}(K) = 2 \end{cases}$$

and let $\rho = \gamma - \gamma^\sigma$. Thus

(10.13) $$\rho = \begin{cases} 2\gamma & \text{if char}(K) \neq 2, \\ 1 & \text{if char}(K) = 2. \end{cases}$$

Let $\bar{v}_1 = v_1$ and for each non-empty $x = i_1 i_2 \cdots i_m \in I$, let $\bar{v}_x$ denote the product $v_{i_m} \cdots v_{i_2} v_{i_1}$ in $C(q,1)$. By 10.9, $\bar{v}_x = \pm v_x$ for all $x \in I$. If $x, y \in I$, then we write

$$x < y$$

whenever x is a subset of y and y contains exactly one more element than x. If $x < y$, then the product $\bar{v}_x v_y$ in $C(q,1)$ is of the form $t v_i$, where t is $\pm \alpha_x$ and i is the unique element in y not in x. If $x = y$, then the product $\bar{v}_x v_y$ in $C(q,1)$ is of the form $\pm \alpha_x v_1$. Thus in both cases we can interpret the product $\bar{v}_x v_y$ as an element of L.

Definition 10.14. Suppose that $\ell = 6$ or 7. Let h be the unique map from $X \times X$ to L which is bilinear over K such that

$$h(r v_x, s v_y) = \begin{cases} (\rho r^\tau s)^{\tau^{|x|}} \bar{v}_x v_y & \text{if } x = y, \\ \rho r^\tau s \bar{v}_x v_y & \text{if } x < y, \\ \rho s^\tau r \bar{v}_y v_x & \text{if } y < x, \\ 0 & \text{otherwise} \end{cases}$$

for all $r, s \in E$ and all $x, y \in I$, where ρ is as in 10.13.

Now suppose that $\ell = 8$ and let $x, y \in I$. We write $x * y$ whenever x and y are disjoint and both of cardinality two. Suppose that $x * y$. Then the product

$$v_{23456} v_x v_y$$

in $C(q,1)$ is of the form $t v_i$, where $t = \pm \alpha_{x \cup y}$ and i is the unique element of $[2,6]$ not in $x \cup y$. We can thus interpret this product as an element of L. We denote this element of L by $v_x * v_y$.

Definition 10.15. Suppose that $\ell = 8$. Let h be the unique map from $X \times X$ to L which is bilinear over K such that

$$h(r v_x, s v_y) = \begin{cases} (\rho r^\tau s)^{\tau^{|x|}} \bar{v}_x v_y & \text{if } x = y, \\ \rho r^\tau s \bar{v}_x v_y & \text{if } x < y, \\ \rho s^\tau r \bar{v}_y v_x & \text{if } y < x, \\ (\rho r s)^\tau v_x * v_y & \text{if } x * y, \\ 0 & \text{otherwise} \end{cases}$$

for all $r, s \in E$ and all $x, y \in I_2$, where ρ is as in 10.13.

We have now reached the goal of the first part of this chapter.

Proposition 10.16. *There exists a map θ from $X \times L$ to L such that*

$$(K, L, q, 1, X, \cdot, h, \theta)$$

is a quadrangular algebra.

Proof. Suppose that $\text{char}(K) \neq 2$. In this case we define θ by the formula in 4.5.i. By 4.8,

$$(K, L, q, 1, X, \cdot, h, \theta)$$

satisfies C1–C4. By 10.10, 10.14 and 10.15, in fact, we have

$$(10.17) \qquad\qquad \theta(tv_1, ev_p) = N(t)\gamma ev_p$$

for all $t, e \in E$ and all $p \in [1, d]$;

$$(10.18) \qquad \theta(tv_i, ev_p) = \begin{cases} -N(t)\alpha_i \gamma ev_p & \text{if } p = 1 \text{ or } p = i, \\ -N(t)\alpha_i \gamma^\sigma ev_p & \text{if } p \notin \{1, i\} \end{cases}$$

for all $i \in [2, d]$, all $t, e \in E$ and all $p \in [1, d]$;

$$(10.19) \qquad \theta(tv_{ij}, ev_p) = \begin{cases} N(t)\alpha_{ij} \gamma ev_p & \text{if } p = 1 \text{ or } p \notin \{i, j\}, \\ N(t)\alpha_{ij} \gamma^\sigma ev_p & \text{if } p \in \{i, j\} \end{cases}$$

for all distinct pairs i, j in $[2, d]$, all $t, e \in E$ and all $p \in [1, d]$; and (only for $d = 4$)

$$(10.20) \qquad\qquad \theta(tv_{234}, ev_p) = -N(t)\alpha_{234}\gamma ev_p$$

for all $t, e \in E$ and all $p \in [1, d]$.

Now suppose that $\text{char}(K) = 2$. Choose an arbitrary ordering $b_1, b_2, \ldots, b_k$ (where $k = 2^{\ell-4}$) of the elements of B; let $\theta(tb, ev_p)$ be as in 10.17–10.20 for all $t, e \in E$, all $b \in B$ and all $p \in [1, d]$; let

$$\theta\left(tb, \sum_p e_p v_p\right) = \sum_p \theta(tb, ev_p)$$

for all $t, e_1, \ldots, e_d \in E$ and all $b \in B$; and let

$$\theta\left(\sum_i t_i b_i, v\right) = \sum_i \theta(t_i b_i, v) + \sum_{i<j} h(t_j b_j, t_i b_i v)$$

for all $t_1, \ldots, t_k \in E$ and all $v \in L$. By (13.35), (13.36) and (13.64) of [12],

$$(K, L, q, 1, X, \cdot, h, \theta)$$

satisfies C1–C4.

Now let

$$\Xi = (K, L, q, 1, X, \cdot, h, \theta)$$

and suppose that $\text{char}(K)$ is arbitrary. By (13.21) of [12], Ξ satisfies B2 and B3; B1 holds by definition. By (13.39) of [12], Ξ satisfies D1 and by (13.49) of [12], Ξ satisfies D2. $\qquad\square$

The proofs of (13.39) and (13.64) in [12], that is, the verifications in the proof of 10.16 that C4 and D1 hold, involve a considerable amount of calculation. It would be desirable to develop a more direct way to see that these two axioms hold.

This concludes the proof of 10.1 in the case that the quadratic space (K, L, q) is of type E_6, E_7 or E_8.

Suppose now that (K, L, q) is a quadratic space of type F_4. Our proof that Ξ exists in this case also involves a certain amount of calculation, but these calculations are shorter and easier to carry out than in the previous case.

By 2.15, we can suppose that

$$L = E \oplus E \oplus F,$$

where E is a field such that

  (i) E/K is a separable quadratic extension;
  (ii) F is a subfield of K such that $K^2 \subset F \subset K$ and scalar multiplication is given by

(10.21) $(u, v, s)t = (ut, vt, st^2)$

     for all $(u, v, s) \in L$ and all $t \in K$;
  (iii) $q(u, v, s) = \beta^{-1}\big(N(u) + \alpha N(v)\big) + s$ for all $(u, v, s) \in L$, where N is the norm of the extension E/K, α is an element of F^* and β is an element of K^*; and
  (iv) q is anisotropic (and hence

(10.22) $\beta \notin F$

     since otherwise $q(1, 0, \beta^{-1}) = 0$).

Let

(10.23) $1 = (0, 0, 1)$.

Thus 1 is in the radical of f and hence the standard involution of the pointed quadratic space $(K, L, q, 1)$ is trivial. Let

$$W = \{(u, v, s) \in L \mid s = 0\}$$

and

$$R = \{(u, v, s) \in L \mid u = v = 0\};$$

thus $L = W \oplus R$.

Let D denote the composition of the fields E^2 and F. Then D/F is a separable quadratic extension and the unique non-trivial element of $\mathrm{Gal}(E/K)$ (which we will now denote by $u \mapsto \bar{u}$) restricts to the unique non-trivial element of $\mathrm{Gal}(D/F)$. Thus the norm of the extension D/F is the restriction of N to D. By 10.22, we have

(10.24) $\beta \notin D$

since $\beta \in K$ is fixed by the map $u \mapsto \bar{u}$.

Let

$$X = D \oplus D.$$

Thus X is both a vector space of dimension four over F (with scalar multiplication denoted by juxtaposition) and a vector space over K with scalar multiplication $*$ given by

(10.25) $$(x, y) * t = (xt^2, yt^2)$$

for all $(x, y) \in X$ and all $t \in K$.

We identify each element $(u, v, 0)$ of W with its image (u, v) in $E \oplus E$. Let q_1 denote the quadratic form on W given by

(10.26) $$q_1(u, v) = \beta^{-1}\big(N(u) + \alpha N(v)\big)$$

for all $(u, v) \in W$ and let f_1 denote the corresponding bilinear form. Thus

(10.27) $$q(b, s) = q_1(b) + s$$

for all $b \in W$ and all $s \in F$ and

(10.28) $$f((b, s), (b', s')) = f_1(b, b')$$

for all $(b, s), (b', s') \in W \oplus R$.

Let q_2 denote the quadratic form on X as a vector space over F given by

(10.29) $$q_2(x, y) = \alpha\big(N(x) + \beta^2 N(y)\big)$$

for all $(x, y) \in X$ and let f_2 be the corresponding bilinear form. Let h denote the map from $X \times X$ to L given by

(10.30) $$h(a, a') = (0, f_2(a, a')) \in W \oplus R = L$$

for all $a, a' \in X$.

Let

(10.31) $$a \cdot (b, s) = \Upsilon(a, b) + sa$$

for all $a \in X$ and all $(b, s) \in W \oplus R = L$, where Υ is the map from $X \oplus W$ to X given by

$$\Upsilon((x, y), (u, v)) = (y\bar{u}^2 + \alpha\bar{y}v^2, \beta^{-2}(xu^2 + \alpha\bar{x}v^2))$$

for all $(x, y) \in D \oplus D = X$ and all $(u, v) \in E \oplus E = W \subset L$.

Let

(10.32) $$\theta(a, (b, s)) = \big(\Theta(a, b), q_2(a)s + \psi(a, b)\big)$$

for all $a \in X$ and all $(b, s) \in L$, where Θ and ψ are maps from $X \oplus W$ to W and from $X \oplus W$ to F given by

$$\Theta((x, y), (u, v)) = (\alpha(\bar{x}v + \beta y\bar{v}), xu + \beta y\bar{u})$$

and

$$\psi((x, y), (u, v)) = \alpha(x\bar{y}u^2 + \bar{x}y\bar{u}^2 + \alpha(xy\bar{v}^2 + \bar{x}\bar{y}v^2))$$

for all $(x, y) \in D \oplus D = X$ and all $(u, v) \in E \oplus E = W$. Let ν denote the map from $X \oplus W$ to K given by

(10.33) $$\nu((x, y), (u, v)) = \alpha(\beta^{-1}(xu\bar{v} + \bar{x}\bar{u}v) + y\bar{u}\bar{v} + \bar{y}uv)$$

for all $(x, y) \in D \oplus D = X$ and all $(u, v) \in E \oplus E = W$.

Proposition 10.34. *The following identities hold for all* $a, a' \in X$ *and* $b, b' \in W$:

(i) $\nu(a, b + b') = \nu(a, b) + \nu(a, b') + f_1(\Theta(a, b), b')$ *and hence*
$f_1(\Theta(a, b), b) = 0$;

(ii) $\psi(a + a', b) = \psi(a, b) + \psi(a', b) + f_2(\Upsilon(a, b), a')$ *and hence*
$f_2(\Upsilon(a, b), a) = 0$;

(iii) $\Upsilon(\Upsilon(a, b), b) = q_1(b)^2 a$;

(iv) $\Theta(a, \Theta(a, b)) = q_2(a)b$;

(v) $\Theta(\Upsilon(a, b), b) + b\nu(a, b) = q_1(b)\Theta(a, b)$;

(vi) $\Upsilon(a, \Theta(a, b)) + a\psi(a, b) = q_2(a)\Upsilon(a, b)$;

(vii) $\nu(\Upsilon(a, b), b) = q_1(b)\nu(a, b)$;

(viii) $\psi(a, \Theta(a, b)) = q_2(a)\psi(a, b)$;

(ix) $\psi(\Upsilon(a, b), b) = q_1(b)^2\psi(a, b)$;

(x) $\nu(a, \Theta(a, b)) = q_2(a)\nu(a, b)$;

(xi) $q_1(\Theta(a, b)) = q_1(b)q_2(a) + \psi(a, b)$;

(xii) $q_2(\Upsilon(a, b)) = q_1(b)^2 q_2(a) + \nu(a, b)^2$.

Proof. These identities follow immediately from the definitions of the various functions involved. $\square$

Proposition 10.35.
$$\Theta(\Upsilon(a, b), b') = \Theta(a, b')q_1(b) + \Theta(a, b)f_1(b, b')$$
$$+ bf_1(\Theta(a, b), b') + b'\nu(a, b)$$

for all $a \in X$ *and all* $b, b' \in W$.

Proof. Choose $a \in X$, $b, b' \in W$ and $t \in K$. By 10.34.v, we have

(10.36)
$$\Theta(\Upsilon(a, b + tb'), b + tb') + (b + tb')\nu(a, b + tb')$$
$$= q_1(b + tb')\Theta(a, b + tb'),$$

where ν is as in 10.33 above. We regard each of these terms as a polynomial in the variable t. The t-coefficient of $\Theta(\Upsilon(a, b + tb'), b + tb')$ is $\Theta(\Upsilon(a, b), b')$, the t-coefficient of $(b + tb')\nu(a, b + tb')$ is, by 10.34.i,

$$b'\nu(a, b) + bf_1(\Theta(a, b), b')$$

and the t-coefficient of $q_1(b + tb')\Theta(a, b + tb')$ is

$$q_1(b)\Theta(a, b') + f_1(b, b')\Theta(a, b).$$

By 2.11, $|K| = \infty$. Hence by [12, (2.26)], the t-coefficient of the left-hand side of 10.36 equals the sum of the t-coefficients of the terms on the right-hand side. $\square$

Proposition 10.37. $\psi(\Upsilon((a, b), b')) = \psi(a, b')q_1(b)^2 + \psi(a, b)f_1(b, b')^2$ *for all* $a \in X$ *and all* $b, b' \in W$.

Proof. Choose $a \in X$, $b, b' \in W$ and $t \in K$. By 10.34.ix, we have

(10.38) $\qquad \psi(\Upsilon(a, b + tb'), b + tb') = q_1(b + tb')^2\psi(a, b + tb')$.

As in the previous proof, we regard each side of 10.38 as a polynomial in t. We have

$$\Upsilon(A, b + tb') = \Upsilon(A, b) + t^2 \Upsilon(A, b')$$

and

$$\psi(A, b + tb') = \psi(A, b) + t^2 \psi(A, b')$$

for all $A \in X$. By 10.34.iii, $\Upsilon(\Upsilon(a, b), b) = q_1(b)^2 a$. By 10.34.ii,

$$f_2(\Upsilon(a, b'), a) = 0.$$

By another application of 10.34.ii, therefore, the t^2-coefficient on the left-hand side of 10.38 is

$$\psi(\Upsilon(a, b), b') + f_2\big(\Upsilon(\Upsilon(a, b), b), \Upsilon(a, b')\big)$$
$$= \psi(\Upsilon(a, b), b') + f_2(\Upsilon(a, b'), a)q_1(b)^2$$
$$= \psi(\Upsilon(a, b), b').$$

The t^2-coefficient of the right-hand side of 10.38 is

$$\psi(a, b')q_1(b)^2 + \psi(a, b)f_1(b, b')^2.$$

By 2.11 and [12, (2.26)], these two coefficients are equal. □

Proposition 10.39. *Let*

$$\Xi = (K, L, q, 1, X, \cdot, h, \theta),$$

where h, $\cdot$ and θ are as in 10.30, 10.31 and 10.32. Then Ξ is a quadrangular algebra. If g and ϕ are as in C3 and C4, then g is identically zero and $\phi(a, (b, s)) = \nu(a, b)$ for all $a \in X$ and all $(b, s) \in W \oplus R$, where ν is as in 10.33.

Proof. By 10.25 and 10.31, the map $(a, v) \mapsto a \cdot v$ is bi-additive and linear over K in the first variable. Moreover,

$$a \cdot \big((b, s)t\big) = a \cdot (bt, st^2) \qquad \text{by 10.21}$$
$$= \Upsilon(a, bt) + sat^2 \qquad \text{by 10.31}$$
$$= \Upsilon(a, b)t^2 + sat^2 = (a \cdot (b, s)) * t \quad \text{by 10.25}$$

for all $a \in X$, all $(b, s) \in W \oplus R = L$ and all $t \in K$;

$$a \cdot 1 = a \cdot (0, 1) = a$$

for all $a \in X$ by 10.23; and

$$\big(a \cdot (b, s)\big) \cdot (b, s) = \Upsilon\big(\Upsilon(a, b) + sa, b\big) + s\big(\Upsilon(a, b) + sa\big)$$
$$= \Upsilon\big(\Upsilon(a, b), b\big) + s^2 a$$
$$= a * q(b, s)$$

by 10.27 and 10.34.iii. Thus Ξ satisfies A1–A3.

The map h (given in 10.30) is symmetric and bi-additive. Moreover,

$$
\begin{aligned}
h(a * t, a') &= \big(0, f_2(a * t, a')\big) \\
&= \big(0, f_2(at^2, a')\big) \quad \text{by 10.25} \\
&= \big(0, f_2(a, a')t^2\big) \\
&= \big(0, f_2(a, a')\big)t \quad \text{by 10.21} \\
&= h(a, a')t
\end{aligned}
$$

for all $a, a' \in X$ and all $t \in K$ and

$$
\begin{aligned}
h(a, a' \cdot (b, s)) &= h(a, \Upsilon(a', b) + sa') \quad &\text{by 10.31} \\
&= (0, \psi(a + a') + \psi(a) + \psi(a') + f_2(a, a')s) \quad &\text{by 10.34.ii,}
\end{aligned}
$$

and thus $h(a, a' \cdot (b, s)) = h(a', a \cdot (b, s))$ for all $a, a' \in X$ and all $(b, s) \in W \oplus R$. By 10.28, we have

$$
f(h(a, b), v) = 0
$$

for all $a, b \in X$ and all $v \in L$. Thus Ξ satisfies B1–B3.

By 10.21 and 10.32, θ is linear over K in the second variable. We have

$$
\begin{aligned}
\theta(a * t, (b, s)) &= (\Theta(at^2, (b, s)), q_2(at^2) + \psi(at^2, b)) \quad &\text{by 10.25} \\
&= (\Theta(a, (b, s))t^2, (q_2(a) + \psi(a, b))t^4) \\
&= \theta(a, (b, s))t^2 \quad &\text{by 10.21}
\end{aligned}
$$

for all $a \in X$, all $(b, s) \in W \oplus R$ and all $t \in K$. Since Θ is additive in the first variable, we have

$$
\begin{aligned}
&\theta(a + a', (b, s)) + \theta(a, (b, s)) + \theta(a', (b, s)) \\
&= (0, f_2(a, a')s + \psi(a + a', b) + \psi(a, b) + \psi(a', b)) \quad &\text{by 10.32} \\
&= (0, f_2(a, a')s + f_2(a, \Upsilon(a', b))) \quad &\text{by 10.34.ii} \\
&= (0, f_2(a, a' \cdot (b, s)) \quad &\text{by 10.31} \\
&= h(a, a' \cdot (b, s)) \quad &\text{by 10.30}
\end{aligned}
$$

for all $a, a' \in X$ and all $(b, s) \in W \oplus R$. Thus Ξ satisfies axioms C1–C3.

Choose $a = (x, y) \in D \oplus D = X$ and $(b, s) \in W \oplus R$. Then

$$
\begin{aligned}
a \cdot \theta(a, (b, s)) &= a \cdot \big(\Theta(a, b), q_2(a)s + \psi(a, b)\big) \quad &\text{by 10.32} \\
&= \Upsilon(a, \Theta(a, b)) + a\big(q_2(a)s + \psi(a, b)\big) \quad &\text{by 10.31} \\
&= \Upsilon(a, b)q_2(a) + asq_2(a) \quad &\text{by 10.34.vi} \\
&= a \cdot (0, q_2(a)) \cdot (b, s) \quad &\text{by 10.31} \\
&= a \cdot \Theta(a, 1) \cdot (b, s) \quad &\text{by 10.23 and 10.32,}
\end{aligned}
$$

so Ξ satisfies D1. Suppose that $\theta(a, (0, 1)) = (0, 1)t$ for some $t \in K^*$. Then

$$
q_2(a) = t^2
$$

by 10.21 and 10.32 and

$$
\begin{aligned}
\alpha\beta q(x + \beta y, \alpha^{-1} t, x\bar{y} + y\bar{x}) & \\
&= \alpha\beta\big(q_1(x + \beta y, \alpha^{-1} t) + x\bar{y} + y\bar{x}\big) && \text{by 10.27} \\
&= \alpha N(x + \beta y) + N(t) + \alpha\beta(x\bar{y} + y\bar{x}) && \text{by 10.26} \\
&= \alpha N(x) + \alpha\beta^2 N(y) + t^2 \\
&= q_2(x, y) + t^2 = q_2(a) + t^2 = 0 && \text{by 10.29.}
\end{aligned}
$$

Since q is anisotropic, it follows that $x + \beta y = 0$. By 10.24, it follows that $x = y = 0$ and hence $a = 0$. We conclude that Ξ satisfies D2.

It remains only to show that Ξ satisfies C4. Choose elements $a \in X$ and $(b, s), (b', s') \in W \oplus R$. Then

$$
\theta(a(b, s), (b', s')) = \Big(\Theta(\Upsilon(a, b) + sa, b'),
$$

$$
q_2(\Upsilon(a, b) + sa)s' + \psi(\Upsilon(a, b) + sa, b')\Big);
$$

$$
\theta(a, (b', s'))q(b, s) = \Big(\Theta(a, b'), q_2(a)s' + \psi(a, b')\Big)(q_1(b) + s);
$$

$$
f((b, s), (b', s'))\theta(a, (b, s)) = \Big(\Theta(a, b), q_2(a)s + \psi(a, b)\Big)f_1(b, b');
$$

and

$$
f\big(\theta(a, (b, s)), (b', s')\big)(b, s) = (b, s)f_1(\Theta(a, b), b').
$$

Since σ is the identity map, it will suffice to show that these four elements add up to something of the form $(b', s')t$ with $t \in K$.

Let B denote the sum of the four first coordinates and let S denote the sum of the four second coordinates. Then

(10.40)
$$
\begin{aligned}
B = {} & \Theta(\Upsilon(a, b), b') + \Theta(a, b')q_1(b) \\
& + \Theta(a, b)f_1(b, b') + bf_1(\Theta(a, b), b')
\end{aligned}
$$

and

(10.41)
$$
\begin{aligned}
S = {} & q_2(\Upsilon(a, b) + sa)s' + \psi(\Upsilon(a, b) + sa, b') \\
& + (q_2(a)s' + \psi(a, b'))(q_1(b) + s)^2 \\
& + (q_2(a)s + \psi(a, b))f_1(b, b')^2 + sf_1(\Theta(a, b), b')^2.
\end{aligned}
$$

By 10.35, $B = b'\nu(a, b)$. We have

$$
\begin{aligned}
q_2(\Upsilon(a, b) + sa) &= q_2(\Upsilon(a, b)) + f_2(\Upsilon(a, b), a)s + q_2(a)s^2 \\
&= q_2(a)\big(q_1(b)^2 + s^2\big) + \nu(a, b)^2;
\end{aligned}
$$

and

$$
\begin{aligned}
\psi(\Upsilon(a, b) + sa, b') &= \psi(\Upsilon(a, b), b') + \psi(a, b')s^2 \\
&\quad + f_2\big(\Upsilon(a, b), \Upsilon(a, b')\big)s
\end{aligned}
$$

by 10.34.ii,xii. By 10.34.i,xii, we have

$$
\begin{aligned}
f_2\big(\Upsilon(a, b), \Upsilon(a, b')\big) &= q_2(\Upsilon(a, b + b')) + q_2(\Upsilon(a, b)) + q_2(\Upsilon(a, b')) \\
&= q_2(a)f_1(b, b')^2 + f_1(\Theta(a, b), b')^2.
\end{aligned}
$$

By 10.41, therefore,

$$S = \nu(a,b)^2 s' + \psi(\Upsilon(a,b), b') + \psi(a, b')q_1(b)^2$$
$$+ \psi(a,b)f_1(b, b')^2.$$

Thus $S = s'\nu(a,b)^2$ by 10.37. Hence by 10.21,

$$(B, S) = (b', s')\nu(a, b).$$

Thus C4 holds with $\phi(a, (b, s)) = \nu(a, b)$.   □

With 10.16 and 10.39, the proof of 10.1 is now complete.

Chapter Eleven

Moufang Quadrangles

In this chapter, we describe the connection between quadrangular algebras and Moufang quadrangles. From 11.15 on, we examine the connection between quadrangular algebras and the stabilizer of an apartment in the automorphism group of an exceptional Moufang quadrangle (and continue with this theme in the next chapter).

Let Γ be a generalized n-gon (for some $n \geq 3$).[1] Thus Γ is a bipartite graph of diameter n whose minimal circuits have length $2n$. We will always assume that Γ is *thick*. This means that $|\Gamma_x| \geq 3$ for all vertices x, where Γ_x denotes the set of vertices adjacent to x.

A path of length n in Γ is called a *root* of Γ and a circuit of length $2n$ is called an *apartment*. The *root group* U_α corresponding to a root

$$\alpha = (x_0, x_1, \ldots, x_n)$$

is the pointwise stabilizer in $\mathrm{Aut}(\Gamma)$ of the set

$$\Gamma_{x_1} \cup \Gamma_{x_2} \cup \cdots \cup \Gamma_{x_{n-1}}.$$

The n-gon Γ is called *Moufang* if for each root α, the root group U_α acts transitively on the set of apartments of Γ containing α.

Suppose now that Γ is Moufang and let Σ be an apartment of Γ. We number the vertices of Σ consecutively by the elements of $\mathbb{Z}_{2n}$. For each $i \in \mathbb{Z}_{2n}$, let U_i denote the root group corresponding to the root

$$(i, i+1, \ldots, i+n).$$

We set

$$U_+ = \langle U_1, \ldots, U_n \rangle.$$

Definition 11.1. A *polygonal sequence* is a sequence of groups

$$(U_+, U_1, \ldots, U_n)$$

(for some $n \geq 3$) which arises from a Moufang n-gon in the manner described above.

---

[1] A generalized n-gon is called a generalized *triangle* if $n = 3$, a generalized *quadrangle* if $n = 4$, etc. A generalized n-gon is also called simply a *generalized polygon* if there is no need to specify n. The notion of a generalized n-gon with $n \geq 3$ is equivalent to the notion of an irreducible spherical building of rank two; see, for instance, [13].

For those Moufang n-gons which arise from algebraic groups, the group U_+ is a maximal unipotent subgroup.

Before going on, we fix some basic conventions. Suppose that G is a group. We set

$$a^b = b^{-1}ab$$

and

$$[a, b] = a^{-1}b^{-1}ab$$

(so $a^b = a \cdot [a, b]$ and $b^a = [a, b^{-1}] \cdot b$) for all $a, b \in G$. If A and B are two subgroups G, then $[A, B]$ denotes the subgroup of G generated by the set

$$\{[a, b] \mid a \in A, \ b \in B\}$$

and if $U_1, \ldots, U_n$ is a sequence of subgroups of G, then

$$U_{[i,j]} = \begin{cases} \langle U_k \mid k \in [i, j] \rangle & \text{if } i \leq j, \\ 1 & \text{otherwise} \end{cases}$$

for all $i, j \in [1, n]$.

With the following definition we can give (in 11.8 below) an "intrinsic" characterization of polygonal sequences.

Definition 11.2. A *root group sequence* is a sequence

$$(U_+, U_1, \ldots, U_n),$$

where U_+ is a group, $n \geq 3$ and $U_1, \ldots, U_n$ are non-trivial subgroups of U_+, satisfying the four conditions $\mathcal{M}_1$–$\mathcal{M}_4$:

$\mathcal{M}_1$: $[U_i, U_j] \subset U_{[i+1, j-1]}$ for all $i, j \in [1, n]$ such that $i < j$.
$\mathcal{M}_2$: The product map $(a_1, a_2, \ldots, a_n) \mapsto a_1 a_2 \cdots a_n$ from

$$U_1 \times U_2 \times \cdots \times U_n$$

to U_+ is bijective.

By $\mathcal{M}_1$, U_n normalizes $U_{[1,n-1]}$. For each $a \in U_n$, let $\tilde{a}$ denote the element $x \mapsto x^a$ in $\mathrm{Aut}(U_{[1,n-1]})$ and let $\tilde{U}_n$ denote the image of U_n in $\mathrm{Aut}(U_{[1,n-1]})$ under the homomorphism $a \mapsto \tilde{a}$.

$\mathcal{M}_3$: There exists a subgroup $\tilde{U}_0$ of $\mathrm{Aut}(U_{[1,n-1]})$ such that for each $a \in U_n^*$, there exists a unique element $\mu(a)$ in $\tilde{U}_0^* \tilde{a} \tilde{U}_0^*$ such that

$$U_j^{\mu(a)} = U_{n-j}$$

for all $j \in [1, n-1]$; and for *some* $e \in U_n^*$, also

$$\tilde{U}_0^{\mu(e)} = \tilde{U}_n \quad \text{and} \quad \tilde{U}_n^{\mu(e)} = \tilde{U}_0.$$

By $\mathcal{M}_1$, U_1 normalizes $U_{[2,n]}$. For each $a \in U_1$, let $\tilde{a}$ denote the element $x \mapsto x^a$ in $\mathrm{Aut}(U_{[2,n]})$ and let $\tilde{U}_1$ denote the image of U_1 in $\mathrm{Aut}(U_{[2,n]})$ under the homomorphism $a \mapsto \tilde{a}$.

$\mathcal{M}_4$: There exists a subgroup $\tilde{U}_{n+1}$ of $\mathrm{Aut}(U_{[2,n]})$ such that for each $a \in U_1^*$, there exists a unique element $\mu(a)$ in $\tilde{U}_{n+1}^* a \tilde{U}_{n+1}^*$ such that

$$U_j^{\mu(a)} = U_{n+2-j}$$

for all $j \in [2, n]$; and for *some* $e \in U_1^*$, also

$$\tilde{U}_{n+1}^{\mu(e)} = \tilde{U}_1 \quad \text{and} \quad \tilde{U}_1^{\mu(e)} = \tilde{U}_{n+1}.$$

Definition 11.3. Two root group sequences

$$(U_+, U_1, \ldots, U_n) \text{ and } (\hat{U}_+, \hat{U}_1, \ldots, \hat{U}_{\hat{n}})$$

are isomorphic if $n = \hat{n}$ and there exists an isomorphism (of groups) from U_+ to $\hat{U}_+$ mapping U_i to $\hat{U}_i$ for all $i \in [1, n]$. If $\Omega = (U_+, U_1, \ldots, U_n)$ is a root group sequence, then

$$(U_+, U_n, \ldots, U_1)$$

is also a root group sequence; it is called the root group sequence *opposite* to Ω and is denoted by Ω^{opp}.

Remark 11.4. Let

$$\Omega = (U_+, U_1, \ldots, U_n)$$

be a root group sequence. By $\mathcal{M}_1$, we have

(11.5) $$[U_1, U_2] = [U_2, U_3] = \cdots = [U_{n-1}, U_n] = 1$$

and for all $i, j \in [1, n]$ such that $i + 1 < j$, all $a \in U_i$ and all $b \in U_j$, there exist elements $c_k \in U_k$ for all $k \in [i + 1, j - 1]$ such that

(11.6) $$[a, b] = c_{i+1} \cdots c_{j-1}.$$

By $\mathcal{M}_2$, every element of U_+ can be written uniquely in the form

(11.7) $$d_1 d_2 \cdots d_n,$$

where $d_i \in U_i$ for all $i \in [1, n]$. (In particular, the elements c_k in 11.6 are uniquely determined by the elements a and b.) In fact, an arbitrary product in U_+ of elements from $U_1 \cup U_2 \cup \cdots \cup U_n$ can be brought into the form 11.7 using only the commutator relations 11.5–11.6 and multiplication in the individual subgroups U_i. It follows that if

$$\hat{\Omega} = (\hat{U}_+, \hat{U}_1, \ldots, \hat{U}_n)$$

is another root group sequence with the same value of n and for all $i \in [1, n]$, ξ_i is an isomorphism from U_i to $\hat{U}_i$, then the maps $\xi_1, \ldots, \xi_n$ extend to an isomorphism from Ω to $\hat{\Omega}$ if and only if

$$[\xi_i(a), \xi_j(b)] = \xi_{i+1}(c_{i+1}) \cdots \xi_{j-1}(c_{j-1})$$

for all $i, j \in [1, n]$ such that $i + 1 < j$, all $a \in U_i$ and all $b \in U_j$, where $c_k \in U_k$ are as in 11.5.

The main results of Chapters 2–8 of [12] can be formulated as follows:

Theorem 11.8. *A sequence of groups is polygonal as defined in 11.1 if and only if it is a root group sequence as defined in 11.2. Let $n \geq 3$, let Γ and $\hat{\Gamma}$ be two Moufang n-gons, let Σ and $\hat{\Sigma}$ be apartments whose vertices have been numbered by the elements of $\mathbb{Z}_{2n}$ and let Ω and $\hat{\Omega}$ be the two corresponding polygonal sequences. Then $\Gamma \cong \hat{\Gamma}$ if and only if $\Omega \cong \hat{\Omega}$ or $\Omega \cong \hat{\Omega}^{\mathrm{opp}}$.*

Proof. By (5.5) and (5.6) of [12], a polygonal sequence satisfies $\mathcal{M}_1$ and $\mathcal{M}_2$. By (6.1) of [12], a polygonal sequence satisfies $\mathcal{M}_3$ and $\mathcal{M}_4$. Every polygonal sequence is thus a root group sequence. The converse is proved in (8.11) of [12]. If $\Gamma \cong \hat{\Gamma}$, then $\Omega \cong \hat{\Omega}$ or $\Omega \cong \hat{\Omega}^{\mathrm{opp}}$ by (4.12) of [12]. The converse is proved in (7.5) of [12]. □

Note, in particular, that if Γ is a Moufang n-gon with $n \geq 3$, then by 11.8, the corresponding polygonal sequence is independent, up to opposites and isomorphism, of the choice of the apartment Σ and the numbering of the vertices of Σ used to produce the sequence.

Example 11.9. Let (K, L, q) an anisotropic quadratic space and let f be as in 1.1.i. Let U_1 and U_3 be two groups (written multiplicatively) both isomorphic to the additive group of K and for $i = 1$ and 3, let $t \mapsto x_i(t)$ be a bijection from K to U_i such that

$$x_i(s + t) = x_i(s)x_i(t)$$

for all $s, t \in K$ (i.e. such that the map $t \mapsto x_i(t)$ is an isomorphism from the additive group of K to U_i). Let U_2 and U_4 be two groups (written multiplicatively) both isomorphic to the additive group of L and for $i = 2$ and 4, let $u \mapsto x_i(u)$ be an isomorphism from the additive group of L to U_i. Then by (32.7) of [12] (and 11.4), there is a unique root group sequence

$$\Omega = (U_+, U_1, \ldots, U_4)$$

such that

$$[x_2(u), x_4(v)^{-1}] = x_3(f(u, v)) \text{ and}$$
$$[x_1(t), x_4(v)^{-1}] = x_2(tv) \cdot x_3(tq(v))$$

for all $u, v \in L$ and all $t \in K$ as well as $[U_1, U_3] = 1$ and

$$[U_1, U_2] = [U_2, U_3] = [U_3, U_4] = 1.$$

Let $\omega \in K^*$ and let

$$\hat{\Omega} = (\hat{U}_+, \hat{U}_1, \ldots, \hat{U}_4)$$

denote the root group sequence obtained by the same construction starting with the quadratic space $(K, L, q/\omega)$. Then the maps $x_1(t) \mapsto \hat{x}_1(t)$, $x_3(t) \mapsto \hat{x}_3(t/\omega)$ and $x_i(u) \mapsto \hat{x}_i(u)$ for $i = 2$ and 4 map the relations defining U_+ to the relations defining $\hat{U}_+$ and hence (by 11.4) extend to an isomorphism from Ω to $\hat{\Omega}$. Thus the root group sequence Ω depends, up to isomorphism, only on the similarity class of the quadratic space (K, L, q).

Proposition 11.10. *Let*
$$\Xi = (K, L, q, 1, X, \cdot, h, \theta)$$
be a quadrangular algebra which is δ-standard (as defined in 4.1) for some $\delta \in L$, let g be as in C3 and let $S = S_\Xi$ denote the set $X \times K$ endowed with the multiplication $\cdot$, where
$$(a, t) \cdot (b, s) = (a + b, s + t + g(b, a))$$
for all $a, b \in X$ and all $s, t \in K$. Then S is a group and
$$(a, t)^{-1} = (-a, -t + g(a, a))$$
for all $(a, t) \in S$.

Proof. By 4.4, g is bilinear. It follows that the multiplication $\cdot$ on S is associative, that
$$(a, t) \cdot (0, 0) = (a, t)$$
and that
$$(a, t) \cdot (-a, -t + g(a, a)) = (0, g(-a, a) + g(a, a)) = (0, 0)$$
for all $(a, t) \in S$.   □

Let Ξ and S be as in 11.10. Suppose that $\operatorname{char}(K) \neq 2$. Then
$$
\begin{aligned}
g(a\pi(a), a) &= \tfrac{1}{2}f(h(a\pi(a), a), 1) && \text{by 4.3} \\
&= \tfrac{1}{2}f(h(a, a), \pi(a)) && \text{by B3} \\
&= f(\pi(a), \pi(a)) && \text{by 4.5.i} \\
&= 2q(\pi(a)) \neq 0 && \text{by D2.}
\end{aligned}
$$
Thus g is not identically zero. By 4.4 and 4.5.ii, g is alternating. It follows that the group S is nonabelian. Now suppose that Ξ is regular (as defined in 1.27) and that $\operatorname{char}(K) = 2$. Then
$$
\begin{aligned}
g(a, b) &= f(h(a, b), \delta) && \text{by 4.3} \\
&= f(h(a, b)^\sigma, \delta^\sigma) && \text{by 1.4} \\
&= f(h(b, a), 1 + \delta) && \text{by 3.6 and 4.1} \\
&= f(h(b, a), 1) + g(b, a) && \text{by 4.3}
\end{aligned}
$$
and
$$
\begin{aligned}
f(h(a\delta, a), 1) &= f(h(a, a), \delta) && \text{by B3} \\
&= g(a, a) && \text{by 4.3} \\
&= f(\pi(a), 1) && \text{by 3.16}
\end{aligned}
$$
for all $a, b \in X$. By 6.2, it follows that g is not symmetric. Therefore S is again nonabelian.

On the other hand, if Ξ is defective (as defined in 1.27), then g is identically zero by 7.2 and S is simply the direct sum of X and the additive group of K.

The notion of a quadrangular algebra arose from the proof of (21.12) in Chapters 26–28 of [12]. This theorem can be reformulated in terms of quadrangular algebras as the sum of 3.1–3.3 and the following result.

Theorem 11.11. *Let* Γ *be a Moufang quadrangle (i.e. a Moufang n-gon with* $n = 4$*), let* Σ *be an apartment whose vertices are numbered by* $\mathbb{Z}_8$*, let*

$$(U_+, U_1, \ldots, U_4)$$

be the corresponding root group sequence, let $Y_1 = C_{U_1}(U_3)$*, let* $Y_3 = C_{U_3}(U_1)$ *and let* $Y_+ = \langle Y_1, U_2, Y_3, U_4 \rangle$*. Suppose that* $Y_1 \neq U_1$ *and that*

$$(Y_+, Y_1, U_2, Y_3, U_4)$$

is a root group sequence which is isomorphic to the root group sequence obtained from an anisotropic quadratic space $(\hat{K}, \hat{L}, \hat{q})$ *as described in 11.9. Then*

  (i) there exists a quadrangular algebra

$$\Xi = (K, L, q, 1, X, \cdot, h, \theta)$$

  which is δ*-standard for some* $\delta \in L$ *such that* (K, L, q) *is similar to* $(\hat{K}, \hat{L}, \hat{q})$ *and*

 (ii) there exist isomorphisms $(a, t) \mapsto x_i(a, t)$ *from the group* S *defined in 11.10 to* U_i *mapping* $(0, K)$ *to* Y_i *for* $i = 1$ *and 3 and isomorphisms* $v \mapsto x_i(v)$ *from the additive group of* L *to* U_i *for* $i = 2$ *and 4 such that*

$$[x_1(a, t), x_3(b, s)^{-1}] = x_2(h(a, b)),$$
$$[x_2(u), x_4(v)^{-1}] = x_3(0, f(u, v)) \text{ and}$$
$$[x_1(a, t), x_4(v)^{-1}] = x_2(\theta(a, v) + tv) \cdot x_3(av, tq(v) + \phi(a, v))$$

for all $(a, t), (b, s) \in S$ *and all* $u, v \in L$ *as well as*

$$[U_1, U_2] = [U_2, U_3] = [U_3, U_4] = 1.$$

Proof. This is proved in (26.1)–(26.32) of [12] (where the basepoint 1 is called ϵ). This part of [12] uses only earlier results from Chapters 2–7 and 21. $\square$

The converse of 11.11 is also valid:

Proposition 11.12. *Let*

$$\Xi = (K, L, q, 1, X, \cdot, h, \theta)$$

be a proper quadrangular algebra which is δ*-standard for some* $\delta \in L$*. Then there exists a unique root group sequence*

$$\Omega_\Xi = (U_+, U_1, \ldots, U_4)$$

as described in 11.11.ii.

Proof. Uniqueness holds by 11.4, so we only have to consider existence. Suppose first that Ξ is special. This means that there is a standard anisotropic pseudo-quadratic space

$$\Pi = (L, \sigma, X, h, \pi)$$

(as defined in 1.16) such that (L, σ) is as in 1.12 and Ξ is the result of applying 1.18 to Π. In this case, the map

$$(a, u) \mapsto (a, \pi(a) + u)$$

is an isomorphism from the group S defined in 11.10 to the group T defined in (11.24) of [12] and the maps $x_i(a, u) \mapsto x_i(a, \pi(a) + u)$ for $i = 1$ and 3 and $x_i(u) \mapsto x_i(u)$ for $i = 2$ and 4 carry the commutator relations in 11.11.ii to the commutator relations given in (16.5) of [12] in terms of Π. The existence of Ω_Ξ holds, therefore, by (32.9) of [12].

Next suppose that (K, L, q) is of type E_6, E_7 or E_8. In this case the existence of Ω_Ξ holds by (32.10) of [12]. The proof of (32.10) requires only (13.67) of [12] and various identities which we have proved in Chapters 3 and 4 of this monograph. (We remind the reader here of comment (viii) on page 7.) The proof of (13.67) of [12] rests, in turn, only on identities which we have proved in Chapters 3 and 4.

Suppose, finally, that (K, L, q) is of type F_4. By 10.25–10.32 and 10.39, Ω_Ξ is precisely the root sequence described in (16.7) of [12]. Existence holds, therefore, by (32.11) of [12]. The proof of (32.11) of [12] requires only the identities in 10.34. □

Proposition 11.13. *Let $\Xi = (K, L, q, \ldots)$ and $\hat{\Xi} = (\hat{K}, \hat{L}, \hat{q}, \ldots)$ be two proper quadrangular algebras and suppose that Ξ is δ-standard for some $\delta \in L$ and that $\hat{\Xi}$ is $\hat{\delta}$-standard for some $\hat{\delta} \in \hat{L}$. Let Ω_Ξ and $\Omega_{\hat{\Xi}}$ be the corresponding root group sequences described in 11.12. If Ξ and $\hat{\Xi}$ are isotopic (as defined in 8.7), then $\Omega_\Xi \cong \Omega_{\hat{\Xi}}$.*

Proof. Suppose first that Ξ is equivalent to $\hat{\Xi}$ as defined in 1.22. Since $\hat{\Xi}$ is $\hat{\delta}$-standard, it follows by 1.23 that there exists $\omega \in K^*$ such that $\hat{\Xi}$ is the quadrangular algebra obtained by applying 1.24 to Ξ, ω and p, where

$$p(a) = -\omega f(\pi(a), \hat{\delta})$$

for all $a \in X$ (so $p(ta) = t^2 p(a)$ for all $t \in K$ and all $a \in X$ by C2). Thus

$$\hat{g}(a, b) = \omega g(a, b) + p(a) + p(b) - p(a + b)$$

and

$$\hat{\phi}(a, v) = \omega \phi(a, v) + p(av) - p(a)q(v)$$

for all $a, b \in X$ and all $v \in L$, where $\hat{g}$ and $\hat{\phi}$ are as in C3 and C4 (with $\hat{h}$ and $\hat{\theta}$ in place of h and θ). If char$(K) \neq 2$, then $\delta = \hat{\delta} = 1/2$ by 4.1.i and hence p is identically zero by 4.1.iii. It follows that the map

$$(a, t) \mapsto (a, \omega t - p(a))$$

is an isomorphism from S to the group $\hat{S}$ obtained from applying 11.10 to $\hat{\Xi}$ and (by 11.4) that the maps

$$x_i(a, t) \mapsto x_i(a, \omega t - p(a))$$

for $i = 1$ and 3, $x_2(v) \mapsto x_2(\omega v)$ and $x_4(v) \mapsto x_4(v)$ extend to an isomorphism from Ω_Ξ to $\Omega_{\hat{\Xi}}$ (in all characteristics).

Now suppose that (ψ_0, ψ_1, ψ_2) is an isomorphism from Ξ to $\hat{\Xi}$ (as defined in 1.25). Then

$$\hat{f}(\psi_1(\delta), \hat{1}) = \hat{f}(\psi_1(\delta), \psi_1(1)) = \psi_0(f(\delta, 1)) = 1.$$

By 4.2 and the conclusion of the previous paragraph, we can thus assume that $\hat{\delta} = \psi_1(\delta)$ and

$$\psi_1(\pi(a)) \equiv \hat{\pi}(\psi_2(a)) \ (\mathrm{mod} \ \langle \hat{1} \rangle)$$

for all $a \in X$ (i.e. that the constant $\hat{\omega}$ in 1.25 equals 1). Since

$$\hat{f}(\psi_1(\pi(a)), \hat{\delta}) = \psi_0(f(\pi(a), \delta)) = 0$$

for all $a \in X$, it follows that, in fact,

$$\psi_1(\pi(a)) = \hat{\pi}(\psi_2(a))$$

for all $a \in X$. By D1, therefore,

$$\begin{aligned}
\psi_2(a)\psi_1(\theta(a, v)) &= \psi_2(a\theta(a, v)) \\
&= \psi_2(a\pi(a)v) \\
&= \psi_2(a)\psi_1(\pi(a))\psi_1(v) \\
&= \psi_2(a)\hat{\pi}(\psi_2(a))\psi_1(v) \\
&= \psi_2(a)\hat{\theta}(\psi_2(a), \psi_1(v))
\end{aligned}$$

and hence, by 3.4 and 3.12,

$$\psi_1(\theta(a, v)) = \hat{\theta}(\psi_2(a), \psi_1(v))$$

for all $a \in X$ and all $v \in L$. By 11.4, we conclude that the maps

$$x_i(a, t) \mapsto x_i(\psi_2(a), \psi_0(t))$$

for $i = 1$ and 3 and $x_i(v) \mapsto x_i(\psi_1(v))$ for $i = 2$ and 4 extend to an isomorphism from Ω_Ξ to $\Omega_{\hat{\Xi}}$.

Finally, let $u \in L^*$ and suppose that $\hat{\Xi}$ is the quadrangular algebra obtained by applying 8.1 to Ξ and u (i.e. that $\hat{\Xi} = \Xi_u$). By 8.1, we have $\hat{g}(a, b) = g(a, b)/q(u)$ and

$$\hat{\phi}(a, v) = \big(\phi(av, u^\sigma) + q(u)\phi(a, v)\big)q(u)^{-3},$$

for all $a, b \in X$ and all $v \in L$. Thus the map $(a, t) \mapsto (a, t/q(u))$ and, by 4.18, the map

$$(a, t) \mapsto (au^{-1}, tq(u)^{-2} + \phi(a, u^{-1})q(u)^{-1})$$

are both isomorphisms from S to $\hat{S}$, and (by 11.4) the maps

$$\begin{aligned}
x_1(a, t) &\mapsto x_1(a, t/q(u)), \\
x_2(v) &\mapsto x_2(v/q(u)), \\
x_3(a, t) &\mapsto x_3(au^{-1}, tq(u)^{-2} + \phi(a, u^{-1})q(u)^{-1}) \text{ and} \\
x_4(v) &\mapsto x_4(v)
\end{aligned}$$

extend to an isomorphism from Ω_Ξ to $\Omega_{\hat{\Xi}}$. By 8.7, we conclude that if Ξ and $\hat{\Xi}$ are isotopic, then $\Omega_\Xi \cong \Omega_{\hat{\Xi}}$. $\qquad \square$

Proposition 11.14. *Let Ξ, $\hat{\Xi}$, etc. be as in 11.13. Suppose that Ξ and $\hat{\Xi}$ are exceptional (as defined in 1.28) and that $\Omega_\Xi \cong \Omega_{\hat{\Xi}}$. Then Ξ is isotopic to $\hat{\Xi}$.*

Proof. By (35.8) of [12] and 11.17 below, there exists, for some $\hat{\gamma} \in \hat{K}^*$, an isomorphism (of pointed quadratic spaces) (ψ_0, ψ_1) from $(K, L, q, 1)$ to $(\hat{K}, \hat{L}, \hat{\gamma}\hat{q}, u)$, where $u = \psi_1(1)$. By 3.2 and 3.3, it follows that Ξ is isomorphic to $\hat{\Xi}_u$. □

It can be deduced from (35.10) in [12] that the conclusion of 11.14 holds also if Ξ is special, but only under the assumption that h (which is a skew-hermitian form in this case) is non-degenerate. We mention, too, that it is possible to obtain an alternative proof of 11.14 by modifying the proof of 11.21 below.

The connection between proper quadrangular algebras and Moufang quadrangles described in 11.11 and 11.12 can be extended to include improper quadrangular algebras. We describe this only briefly. Suppose that Ξ is an *improper* quadrangular algebra, i.e. that the bilinear form f is identically zero. Suppose first that h is not identically zero. Replace Ξ by the equivalent quadrangular algebra $\hat{\Xi}$ described in 9.23 and let V, Q and so on be as in 9.24. Then the commutator relations given in 11.11.ii define a root group sequence which is isomorphic to the opposite of the root group sequence defined in (16.3) of [12] in terms of the anisotropic quadratic space (L, V, Q). Suppose, instead, that h is identically zero. Let (F, L_0, X_0) be as in 9.30. Then the commutator relations given in 11.11.ii define a root group sequence which is isomorphic to the opposite of the root group sequence defined in (16.4) of [12] in terms of the indifferent set (F, L_0, X_0). It follows from (38.4) of [12], 9.27 and 9.31 that *all* Moufang quadrangles which are either of indifferent type or of degenerate quadratic form type[2] can be accounted for in this way.

We now turn our attention to an important subgroup of the automorphism group of an exceptional Moufang polygon.

Notation 11.15. Suppose that

$$\Xi = (K, L, q, 1, X, \cdot, h, \theta)$$

is an exceptional quadrangular algebra, that Ξ is δ-standard for some $\delta \in L$ and that

$$\Omega_\Xi = (U_+, U_1, \ldots, U_4)$$

is the root sequence described in 11.12. Let

$$H = \{\tau \in \operatorname{Aut}(U_+) \mid U_i^\tau = U_i \text{ for all } i \in [1, 4]\}.$$

By (7.5) of [12], H is (up to isomorphism) the pointwise stabilizer of an apartment in the automorphism group of the Moufang quadrangle determined by Ω_Ξ. We will study this group for the rest of this chapter; see especially 11.20 and (in the next chapter) 12.11.

---

[2]That is, of quadratic form type defined by an anisotropic quadratic space which is degenerate.

Notation 11.16. Let $i, j, k \in [1, 4]$ with $i < k < j$ and suppose that $a \in U_i$ and $b \in U_j$. By $\mathcal{M}_1$ and $\mathcal{M}_2$, the commutator $[a, b]$ can be written uniquely as a product $c_{i+1} \cdots c_{j-1}$ with $c_m \in U_m$ for each $m \in [i + 1, j - 1]$; let

$$[a, b]_k$$

denote the factor c_k.

Lemma 11.17. $x_1(0, K) = C_{U_1}(U_3)$ and $x_3(0, K) = C_{U_3}(U_1)$.

Proof. Suppose that there exist elements $a \in X^*$ such that $h(a, X) = 0$. By 7.44, 7.49 and 7.50, (K, L, q) is not of type F_4. By 11.15 and 3.1–3.3, therefore, (K, L, q) is of type E_6, E_7 or E_8. By 6.8, X is an irreducible right $C(q, 1)$-module. By B2 and 3.6, $h(b, X) = 0$ for some $b \in X$ implies that $h(bv, X) = 0$ for all $v \in L$. It follows (by B1) that $h(X, X) = 0$. This is impossible, however, by 6.12. We conclude that $h(a, X) \neq 0$ for all $a \in X^*$. By 3.6 again, it follows that $h(X, a) \neq 0$ for all $a \in X^*$. By 11.11.ii, therefore, $x_1(0, K) = C_{U_1}(U_3)$ and $x_3(0, K) = C_{U_3}(U_1)$. $\quad\square$

Proposition 11.18. *For each $\nu \in H$, there exist elements $\omega \in K^*$ and $\tau \in \mathrm{Aut}(K)$ such that*

$$x_1(0, t)^\nu = x_1(0, \omega t^\tau)$$

for all $t \in K$.

Proof. Choose $\nu \in H$. By 11.17, $x_1(0, K)$ is normalized by H. Let λ be the map from K to itself such that

$$x_1(0, t)^\nu = x_1(0, \lambda(t))$$

for all $t \in K$ and let ξ_i (for $i = 2$ and 4) be the maps from L to itself such that

$$x_i(v)^\nu = x_i(\xi_i(v))$$

for all $v \in L$. The maps λ, ξ_2 and ξ_4 are all additive bijections. We have (by 11.16 and 11.11.ii)

$$[x_1(0, t), x_4(v)^{-1}]_2 = x_2(tv)$$

for all $v \in L$ and all $t \in K$. Applying ν to this identity, we find that

(11.19) $$\xi_2(tv) = \lambda(t)\xi_4(v)$$

for all $v \in L$ and all $t \in K$. Setting $t = 1$ in 11.19 and letting $\omega = \lambda(1)$, we have

$$\xi_4(v) = \omega^{-1}\xi_2(v)$$

for all $v \in L$. Therefore

$$\xi_2(tv) = \omega^{-1}\lambda(t)\xi_2(v)$$

for all $v \in L$ and all $t \in K$. Now choose $v \in L^*$. Then

$$\omega^{-1}\lambda(st)\xi_2(v) = \xi_2(stv)$$
$$= \omega^{-1}\lambda(s)\xi_2(tv)$$
$$= \omega^{-2}\lambda(s)\lambda(t)\xi_2(v)$$

and thus
$$\lambda(st) = \omega^{-1}\lambda(s)\lambda(t)$$
for all $s, t \in K$. It follows that the map τ from K to itself given by
$$\tau(t) = \omega^{-1}\lambda(t)$$
for all $t \in K$ is an automorphism of K. $\qquad\qquad\qquad\qquad$ □

Definition 11.20. Let H be as in 11.15. The map $\nu \mapsto \tau$ from H to $\mathrm{Aut}(K)$, where τ is as in 11.18, is a homomorphism. Let H_0 denote its kernel. The elements of H_0 are the K-*linear* elements of H.

If the Moufang quadrangle corresponding to Ω_Ξ is the building coming from a K-form of a simple algebraic group (equivalently, if (K, L, q) is of type E_6, E_7 or E_8), then H_0 is the anisotropic kernel of this group. In 12.11 below, we will give another interpretation of H_0.

Proposition 11.21. *Let $\nu \in H_0$, where H_0 is as in 11.20. Then there exist unique K-linear automorphisms ξ and ψ of L, respectively X, and an element $\omega \in K^*$ such that the following hold:*

(i) $q(\xi(v))/q(u) = q(v)$, where $u = \xi(1)$;
(ii) $\psi(av) = \psi(a)\xi(v)u^{-1}$;
(iii) $\xi(h(a, b)) = h(\psi(a), \psi(b)u)\omega^{-1}$; and
(iv) $\xi(\theta(a, v)) = \big(\theta(\psi(a), \xi(v)) + M(a)\xi(v)\big)\omega^{-1}$, where

$$M(a) = \begin{cases} 0 & \text{if } \mathrm{char}(K) \neq 2, \\ f(\theta(\psi(a), u), \xi(\delta))q(u)^{-1} & \text{if } \mathrm{char}(K) = 2 \end{cases}$$

for all $a, b \in X$ and all $v \in L$; and

(11.22)
$$\begin{aligned} x_1(a, t)^\nu &= x_1(\psi(a), \omega t + M(a)), \\ x_2(v)^\nu &= x_2(\xi(v)\omega), \\ x_3(a, t)^\nu &= x_3\big(\psi(a)u, \omega q(u)t + M(a)q(u) + \phi(\psi(a), u)\big), \\ x_4(v)^\nu &= x_4(\xi(v)) \end{aligned}$$

for all $(a, t) \in S$ and all $v \in L$.

Proof. By 11.20, there exist maps ψ from X to itself and M from S to K and an element $\omega \in K^*$ such that
$$x_1(a, t)^\nu = x_1(\psi(a), \omega t + M(a, t))$$
and $M(0, t) = 0$ for all $(a, t) \in S$. By 11.17, there exist maps $\bar\lambda$ from K to itself, $\bar\psi$ from X to itself and $\bar{M}$ from S to K such that
$$x_3(a, t)^\nu = x_3(\bar\psi(a), \bar\lambda(t) + \bar{M}(a, t))$$
and $\bar{M}(0, t) = 0$ for all $(a, t) \in S$. Both the maps ψ and $\bar\psi$ are bijections and, by 11.10, additive. Let ξ and $\bar\xi$ denote the two maps from L to itself such that
$$x_2(v)^\nu = x_2(\bar\xi(v))$$

and

$$x_4(v)^\nu = x_4(\xi(v))$$

for all $v \in L$. Both ξ and $\bar{\xi}$ are additive bijections from L to itself.

Applying ν to the identity

$$[x_1(a,t), x_4(v)^{-1}]_3 = x_3(av, tq(v) + \phi(a,v)),$$

we find (by 11.11.ii) that

(11.23) $$\psi(a)\xi(v) = \bar{\psi}(av)$$

and

(11.24) $$(\omega t + M(a,t))q(\xi(v)) + \phi(\psi(a), \xi(v)) = \bar{\lambda}(T) + \bar{M}(av, T)$$

for all $(a,t) \in S$ and all $v \in L$, where

$$T = tq(v) + \phi(a,v).$$

Setting $v = 1$ in 11.23, we find (by A2) that

(11.25) $$\bar{\psi}(a) = \psi(a)u$$

for all $a \in X$, where $u = \xi(1)$. Therefore

(11.26) $$\psi(av) = \psi(a)\xi(v)u^{-1}$$

for all $a \in X$ and all $v \in L$ by A3 and 11.23. Thus (ii) holds. By 4.5.iii, 4.13 and 4.14, we have $\phi(a,v) = 0$ if $a = 0$ or $v = 1$. Setting $a = 0$ and $v = 1$ in 11.24, we find that

(11.27) $$\bar{\lambda}(t) = t\omega q(u)$$

for all $t \in K$. Setting $a = 0$ and $t = 1$ in 11.24, therefore, we have

(11.28) $$q(\xi(v)) = q(u)q(v)$$

for all $v \in L$. Thus (i) holds. Substituting 11.27 and 11.28 into 11.24 and then setting $v = 1$, we then obtain (by A2)

$$\bar{M}(a,t) = M(a,t)q(u) + \phi(\psi(a), u)$$

for all $(a,t) \in S$.

Applying ν to the identity

$$[x_1(a,t), x_4(v)^{-1}]_2 = x_2(\theta(a,v) + tv),$$

we obtain

(11.29) $$\bar{\xi}(\theta(a,v) + tv) = \theta(\psi(a), \xi(v)) + (\omega t + M(a,t))\xi(v)$$

for all $(a,t) \in S$ and all $v \in L$. Setting $a = 0$ in 11.29, we have (by 3.12)

(11.30) $$\bar{\xi}(tv) = \omega t\xi(v)$$

for all $t \in K$ and all $v \in L$. Setting $t = 1$ in 11.30, we have

(11.31) $$\bar{\xi}(v) = \omega\xi(v)$$

for all $v \in L$. It follows (by 11.30) that ξ is linear over K. By A1, A2 and 11.26, it follows that ψ is also K-linear. Applying ν to the identity

$$[x_1(a,t), x_3(b,s)^{-1}] = x_2(h(a,b)),$$

we have (by 11.25 and 11.31)

$$\omega\xi(h(a,b)) = h(\psi(a), \psi(b)u)$$

for all $a, b \in X$. Thus (iii) holds.

Since ξ is linear over K, 11.29 and 11.31 imply that

(11.32) $$\omega\xi(\theta(a,v)) = \theta(\psi(a), \xi(v)) + M(a,t)\xi(v)$$

for all $(a,t) \in S$ and all $v \in L$. By 11.32, we conclude that M is independent of t, so we can set $M(a) = M(a,t)$ for all $(a,t) \in S$. By 4.1 and 11.28,

$$f(\xi(\pi(a)), \xi(\delta)) = q(u)f(\pi(a), \delta) = 0$$

for all $a \in X$ and

$$f(\xi(1), \xi(\delta)) = q(u).$$

By 11.32 (with 1 in place of v), we thus have

$$f(\theta(\psi(a), u), \xi(\delta)) + q(u)M(a) = 0.$$

If $\operatorname{char}(K) \neq 2$, then $u = 2\xi(\delta)$ (since $\delta = 1/2$) and $f(\pi(\psi(a)), 1) = 0$ by 4.1 and hence M is identically zero by 4.9.i. Thus (iv) holds. $\square$

Proposition 11.33. *Let ν, ψ and ξ be as in 11.21. Then ν is uniquely determined by ψ and ξ.*

Proof. It suffices to observe that by D2 and 11.21.iv, the constant ω and the map M are both uniquely determined by ψ and ξ. $\square$

Let Γ denote the Moufang quadrangle which corresponds to the root group sequence Ω_Ξ. This means that for each apartment Σ of Γ, the vertices of Σ can be labeled consecutively by the elements of $\mathbb{Z}_8$ so that (up to isomorphism) Ω_Ξ is the polygonal sequence described in 11.1, the groups $\tilde{U}_0$ and $\tilde{U}_5$ in $\mathcal{M}_3$ and $\mathcal{M}_4$ are the roots groups U_0 and U_5 of Γ (as defined just before 11.1) and H is the pointwise stabilizer of Σ in $\operatorname{Aut}(\Gamma)$. Let $G = \operatorname{Aut}^\circ(\Gamma)$ be as defined in (33.3) of [12] and let $G^\dagger$ denote the subgroup of G generated by all the root groups. Then $G^\dagger$ is a normal subgroup of G and, by (37.3) of [12], $G^\dagger$ is a simple group. Let

$$H_i^\dagger = \langle \mu(a)\mu(b) \mid a,b \in U_i^* \rangle$$

for $i = 1$ and 4 and let $H^\dagger = H_1^\dagger H_4^\dagger$. By (6.1) of [12], we have

$$U_5^{\mu(x_1(a,t))} = U_1 \quad \text{and} \quad U_1^{\mu(x_1(a,t))} = U_5$$

for all $(a,t) \in S^*$ and

$$U_0^{\mu(x_4(v))} = U_4 \quad \text{and} \quad U_4^{\mu(x_4(v))} = U_0$$

for all $v \in L^*$. Thus $H^\dagger \subset H$. By (33.9) and (37.8) of [12], $H^\dagger$ is a normal subgroup of the group H and

$$G/G^\dagger \cong H/H^\dagger.$$

The subgroup $H^\dagger$ is generated by the elements $h_1(a,t)$ for $(a,t) \in S^*$ and $h_4(v)$ for $v \in L^*$ defined in 11.35 below. In particular, $H^\dagger$ is contained in the subgroup H_0 defined in 11.20. In [2], it is shown that $H_0 = H^\dagger$ if the quadrangle Γ is of type F_4. In [14], it is shown that

$$H_0/H^\dagger \cong \operatorname{Aut}(E/K) \cdot E^*/(E^*)^3$$

if Γ is of type E_6 and

$$H_0/H^\dagger \cong D^*/[D^*, D^*]K^*$$

if Γ is of type E_7, where E and D are as in 2.24; probably $H_0 = H^\dagger$ if Γ is of type E_8, but to prove this seems to be a difficult problem.

Notation 11.34. For each $(a,t) \in S^*$, let $r = q(\pi(a)+t)$ (so $r \neq 0$ by D2) and let $a' = a(\pi(a)+t)/r$. We then set

$$\zeta_{a,t}(b) = b - ah(a',b),$$
$$\chi(a,b,t) = \phi(a, h(a',b)) + tq(h(a',b)) - g(ah(a',b),b) \text{ and}$$
$$\chi_0(a,b,t) = \chi(a,b,t)q(\pi(a)+t) + \phi(\zeta_{a,t}(b), \pi(a)^\sigma + t)$$

for all $b \in X$.

Proposition 11.35. *Let*

$$h_1(a,t) = \mu(x_1(0,1))^{-1}\mu(x_1(a,t))$$

and

$$h_4(v) = \mu(x_4(1))^{-1}\mu(x_4(v))$$

for all $(a,t) \in S^*$ and all $v \in L^*$. Then

$$x_1(b,s)^{h_1(a,t)} = x_1\big(\zeta_{a,t}(b)(\pi(a)^\sigma + t), sq(\pi(a)+t) + \chi_0(a,b,t)\big),$$
$$x_2(u)^{h_1(a,t)} = x_2(\theta(a,u) + tu),$$
$$x_3(b,s)^{h_1(a,t)} = x_3\big(\zeta_{a,t}(b), s + \chi(a,b,t)\big),$$
$$x_4(u)^{h_1(a,t)} = x_3\big((\theta(a,u) + tu)/q(\pi(a)+t)\big);$$

where $\zeta_{a,t}$, χ and χ_0 are as in 11.34, and

$$x_1(b,s)^{h_4(v)} = x_1\big(bv^{-1}, t/q(v) + \phi(b, v^{-1})\big),$$
$$x_2(u)^{h_4(v)} = x_2(f(u^\sigma, v)v/q(v) - u^\sigma),$$
$$x_3(b,s)^{h_4(v)} = x_3(bv, tq(v) + \phi(b, v)),$$
$$x_4(u)^{h_4(v)} = x_4(vf(v, u^\sigma) - q(v)u^\sigma)$$

for all $(b,s) \in S$, all $u \in L$, all $(a,t) \in S^*$ and all $v \in L^*$. In particular, $h_1(a,t)$ and $h_4(v)$ lie in the subgroup H_0 (defined in 11.20) for all $(a,t) \in S^*$ and all $v \in L^*$.

Proof. Choose $(a, t) \in S^*$ and $v \in L^*$. The formulas for the action of $h_1(a, t)$ on an element of $U_{[2,4]}$ and the formulas for the action of $h_4(v)$ on an element of $U_{[1,3]}$ hold by (32.10) of [12]. It remains, therefore, only to determine how $h_1(a, t)$ acts on U_1 and $h_4(v)$ on U_4.

Choose $(b, s) \in S$ and let (b', s') be the unique element of S such that

$$x_1(b', s') = x_1(b, s)^{h_1(a,t)}.$$

Since $\phi(b, 1) = 0$ (by 4.13), we have (by 11.11.ii)

$$[x_1(b, s), x_4(1)^{-1}]_3 = x_3(b, s).$$

Conjugating by $h_1(a, t)$, we obtain

$$[x_1(b', s'), x_4((\pi(a) + t)/q(\pi(a) + t))^{-1}]_3 = x_3(\zeta_{a,t}(b), s + \chi(a, b, t)).$$

On the other hand,

$$[x_1(b', s'), x_4((\pi(a) + t)/q(\pi(a) + t))^{-1}]_3$$
$$= x_3\Big(b'(\pi(a) + t)/q(\pi(a) + t),$$
$$\qquad s'/q(\pi(a) + t) + \phi(b', (\pi(a) + t)/q(\pi(a) + t))\Big)$$

(by 11.11.ii). Thus

$$b' = \zeta_{a,t}(b)(\pi(a) + t)^{\sigma}$$

by A3 and

$$s' = \big(s + \chi(a, b, t) + \phi(b', (\pi(a) + t)/q(\pi(a) + t))\big)q(\pi(a) + t).$$

By 4.23,

$$\phi(b', (\pi(a) + t)/q(\pi(a) + t)) = \phi(\zeta_{a,t}(b), \pi(a)^{\sigma} + t)/q(\pi(a) + t).$$

Thus

$$s' = sq(\pi(a) + t) + \chi_0(a, b, t),$$

where χ_0 is as defined in 11.34.

Now choose $u \in L$ and let u' be the unique element of L such that

$$x_4(u') = x_4(u)^{h_4(v)}.$$

We have

$$[x_1(0, 1), x_4(u)^{-1}]_2 = x_2(u).$$

Conjugating by $h_4(v)$, we obtain

$$[x_1(0, 1/q(v)), x_4(u')^{-1}]_2 = x_2(f(u^{\sigma}, v)v/q(v) - u^{\sigma}).$$

On the other hand,

$$[x_1(0, 1/q(v)), x_4(u')^{-1}]_2 = x_2(u'/q(v))$$

(by 11.11.ii). Therefore

$$u' = f(u^{\sigma}, v)v - u^{\sigma}q(v).$$

$\square$

Chapter Twelve

The Structure Group

The structure group of a quadrangular algebra is essentially the automorphism group of its isotopy class. In 12.11 below, we show that the structure group of an exceptional quadrangular algebra is isomorphic to the group of "linear" elements in the pointwise stabilizer of an apartment in the automorphism group of the corresponding Moufang quadrangle. (This is the group we called H_0 in Chapter 11; see 11.20.)

Let

$$\Xi = (K, L, q, 1, X, \cdot, h, \theta)$$

be a quadrangular algebra which we assume to be standard (as defined in 4.1) if $\operatorname{char}(K) \neq 2$. Let $[\Xi]$ denote the set

$$\{\Xi_u \mid u \in L^*\},$$

where the Ξ_u are the isotopes of Ξ as defined in 8.6 and 8.7. Thus

$$\Xi_u = (K, L, q/q(u), u, X, \cdot_u, h_u, \theta_u)$$

for all $u \in L^*$, where

(12.1) $$a \cdot_u v = avu^{-1},$$

(12.2) $$h_u(a, b) = h(a, bu)q(u)^{-1},$$

(12.3) $$\theta_u(a, v) = q(u)^{-1}\theta(a, v)$$

for all $a, b \in X$ and all $v \in L$. In particular, the isotopes Ξ_u (for $u \in L^*$) are all quadrangular algebras based on the same two vector spaces L and X over the same field K. By 8.9, $[\Xi_u] = [\Xi]$ for all $u \in L^*$.

Notation 12.4. Let $\operatorname{Str}(\Xi)$ denote the set of all pairs (ξ, ψ), where ξ and ψ are K-linear automorphisms of the vector spaces L and X, respectively, such that (ξ, ψ) is a K-linear isomorphism (as defined in 1.25) from Ξ to $\Xi_{\xi(1)}$.

In the next three results, we show that the elements of $\operatorname{Str}(\Xi)$ form a group permuting the isotopes of Ξ.

Proposition 12.5. *Suppose that ξ and ψ are K-linear automorphisms of L and X, respectively, and let $u = \xi(1)$. Then*

$$(\xi, \psi) \in \operatorname{Str}(\Xi)$$

if and only if

(i) $q(\xi(v))/q(u) = q(v)$ for all $v \in L$;

(ii) $\psi(av) = \psi(a) \cdot_u \xi(v) = \psi(a)\xi(v)u^{-1}$ for all $a \in X$ and all $v \in L$; and

(iii) there exists an element $\omega \in K^*$ such that

(12.6)
$$\xi(h(a,b)) = \omega h_u(\psi(a), \psi(b))$$
$$= \omega h(\psi(a), \psi(b)u)q(u)^{-1};$$

and

(12.7)
$$\xi(\theta(a,v)) \equiv \omega\theta_u(\psi(a), \xi(v))$$
$$\equiv \omega q(u)^{-1}\theta(\psi(a), \xi(v)) \ (\mathrm{mod} \ \langle \xi(v) \rangle)$$

for all $a, b \in X$ and all $v \in L$.

Proof. This holds by 1.25 and 12.4.   □

Proposition 12.8. *Let $(\xi, \psi) \in \mathrm{Str}(\Xi)$. Then (ξ, ψ) is a K-linear isomorphism from Ξ_w to $\Xi_{\xi(w)}$ for all $w \in L^*$.*

Proof. Let $u = \xi(1)$, let ω be as in 12.5.iii and choose $w \in L^*$. By 12.5.i,
$$q(\xi(v))/q(\xi(w)) = q(u)q(v)/q(u)q(w) = q(v)/q(w)$$
for all $v \in L$. Let $b \in X$. Then

$$bu^{-1}\xi(w^{-1})u^{-1} = bu^{-1}f(u^{-1}, \xi(w^{-1})^\sigma)$$
$$- b\xi(w^{-1})^\sigma q(u)^{-1} \qquad \text{by 3.9}$$
$$= bu^\sigma f(u, \xi(w^{-1}))q(u)^{-2}$$
$$- b\xi(w^{-1})^\sigma q(u)^{-1} \qquad \text{by 1.3 and 1.4}$$
$$= bu^\sigma f(\xi(1), \xi(w^{-1}))q(u)^{-2}$$
$$- b\xi(w^{-1})^\sigma q(u)^{-1}$$
$$= b\big(uf(1, w^{-1}) - \xi(w^{-1})\big)^\sigma q(u)^{-1} \qquad \text{by 12.5.i}$$
$$= b\xi(1f(1, w) - w^\sigma)^\sigma q(u)^{-1}q(w)^{-1}$$
$$= b\xi(w)^\sigma q(u)^{-1}q(w)^{-1} = b\xi(w)^{-1} \qquad \text{by 12.5.i.}$$

Therefore

$$\psi(a \cdot_w v) = \psi(avw^{-1}) \qquad \qquad \text{by 12.1}$$
$$= \psi(a)\xi(v)u^{-1}\xi(w^{-1})u^{-1} \qquad \text{by 12.5.ii}$$
$$= \psi(a)\xi(v)\xi(w)^{-1}$$
$$= \psi(a) \cdot_{\xi(w)} \xi(v) \qquad \qquad \text{by 12.1}$$

for all $a \in X$ and all $v \in L$;

$$\xi(h_w(a,b)) = \xi(h(a, bw))q(w)^{-1} \qquad \qquad \text{by 12.2}$$
$$= \omega h(\psi(a), \psi(bw)u)q(u)^{-1}q(w)^{-1} \qquad \text{by 12.6}$$
$$= \omega h(\psi(a), \psi(b)\xi(w))q(u)^{-1}q(w)^{-1} \qquad \text{by A3 and 12.5.ii}$$
$$= \omega h(\psi(a), \psi(b)\xi(w))q(\xi(w))^{-1} \qquad \text{by 12.5.i}$$
$$= \omega h_{\xi(w)}(\psi(a), \psi(b)) \qquad \qquad \text{by 12.2}$$

for all $a, b \in X$; and

$$\begin{aligned}
\xi(\theta_w(a,v)) &\equiv q(w)^{-1}\xi(\theta(a,v)) && \text{by 12.3} \\
&\equiv wq(u)^{-1}q(w)^{-1}\theta(\psi(a),\xi(v)) && \text{by 12.7} \\
&\equiv wq(\xi(w))^{-1}\theta(\psi(a),\xi(v)) && \text{by 12.5.i} \\
&\equiv w\theta_{\xi(w)}(\psi(a),\xi(v)) \ (\mathrm{mod}\ \langle\xi(v)\rangle) && \text{by 12.3}
\end{aligned}$$

for all $a \in X$ and all $v \in L$. By 1.25, therefore, the pair (ξ,ψ) is an isomorphism from Ξ_w to $\Xi_{\xi(w)}$. $\qquad\square$

Theorem 12.9. *The set* $\mathrm{Str}(\Xi)$ *is a group with multiplication given by*

$$(\xi_1,\psi_1)\cdot(\xi_2,\psi_2) = (\xi_1\xi_2,\psi_1\psi_2)$$

for all $(\xi_1,\psi_1),(\xi_2,\psi_2) \in \mathrm{Str}(\Xi)$.

Proof. By 12.8, the set $\mathrm{Str}(\Xi)$ is closed under multiplication. We thus only need to show that it is also closed under inverses. Choose $(\xi,\psi) \in \mathrm{Str}(\Xi)$ and let $u = \xi(1)$. Let $a, b \in X$ and $v \in L$. By 12.5.i,

$$q(\xi^{-1}(v)) = q(u)^{-1}q(v)$$

for all $v \in L$. In particular, $q(\xi^{-1}(1)) = q(u)^{-1}$ and thus

$$q(\xi^{-1}(v))/q(\xi^{-1}(1)) = q(v)$$

for all $v \in L$. By A2 and 12.5.ii, we obtain avu^{-1} whether we apply ψ to $\psi^{-1}(av)\xi^{-1}(1)$ or to $\psi^{-1}(a)\xi^{-1}(v)$. Since ψ is injective, it follows (by A3) that

$$\begin{aligned}
\psi^{-1}(av) &= \psi^{-1}(a)\xi^{-1}(v)\xi^{-1}(1)^{-1} \\
&= \psi^{-1}(a) \cdot_{\xi^{-1}(1)} \xi^{-1}(v).
\end{aligned}$$

By 12.8, there exists an element $w \in K^*$ such that

$$\xi\Big(h_{\xi^{-1}(1)}\big(\psi^{-1}(a),\psi^{-1}(b)\big)\Big) = wh(a,b)$$

and

$$\xi\Big(\theta_{\xi^{-1}(1)}\big(\psi^{-1}(a),\xi^{-1}(v)\big)\Big) \equiv w\theta(a,v) \ (\mathrm{mod}\ \langle v\rangle)$$

for all $a, b \in X$ and all $v \in L$ (since $h_1 = h$ and $\theta_1 = \theta$). Thus the pair (ξ^{-1},ψ^{-1}) is an isomorphism from Ξ to $\Xi_{\xi^{-1}(1)}$. $\qquad\square$

Definition 12.10. The group $\mathrm{Str}(\Xi)$ described in 12.4 and 12.9 is called the *structure group* of Ξ. By 12.8, $\mathrm{Str}(\Xi)$ depends only on the isotopy class $[\Xi]$ of Ξ.

Note that the group of K-linear automorphisms of Ξ is the stabilizer in $\mathrm{Str}(\Xi)$ of the basepoint 1.

Here is our final result.

Theorem 12.11. *Suppose* Ξ *is exceptional and δ-standard for some $\delta \in L$. Then* $\mathrm{Str}(\Xi) \cong H_0$*, where H_0 is the group defined in 11.20.*

Proof. Let $\nu \in H_0$ and let (ξ, ψ) be as in 11.21. By 12.5,

$$(\xi, \psi) \in \mathrm{Str}(\Xi),$$

and by 11.33 and 12.9, the map

$$\nu \mapsto (\xi, \psi)$$

is an injective homomorphism from H_0 to $\mathrm{Str}(\Xi)$. We thus only need to show that this map is surjective.

Choose $(\xi, \psi) \in \mathrm{Str}(\Xi)$ and let $u = \xi(1)$. By 12.7, there exist a map M from $X \times L$ to K and an element $\omega \in K^*$ such that

$$(12.12) \qquad \xi(\theta(a, v)) = \omega q(u)^{-1} \big(\theta(\psi(a), \xi(v)) + M(a, v)\xi(v) \big)$$

for all $a \in X$ and all $v \in L$. Then

$$\psi(a) \Big(\omega q(u)^{-1}(\theta(\psi(a), \xi(v)) + M(a, v)\xi(v)) \Big) u^{-1}$$

$$\begin{aligned}
&= \psi(a)\xi(\theta(a, v))u^{-1} && \text{by 12.12} \\
&= \psi(a\theta(a, v)) && \text{by 12.5.ii} \\
&= \psi(a\pi(a)v) && \text{by D1} \\
&= \psi(a)\xi(\pi(a))u^{-1}\xi(v)u^{-1} && \text{by 12.5.ii}
\end{aligned}$$

$$= \psi(a) \Big(\omega q(u)^{-1}(\theta(\psi(a), u) + M(a, 1)u) \Big) u^{-1}\xi(v)u^{-1} \quad \text{by 12.12}$$

for all $a \in X$ and all $v \in L$. By A1, A3 and D1, we have

$$\psi(a)\omega q(u)^{-1}\theta(\psi(a), \xi(v))u^{-1}$$

$$\begin{aligned}
&= \psi(a)\omega q(u)^{-1}\pi(\psi(a))\xi(v)u^{-1} \\
&= \psi(a)\omega q(u)^{-1}\pi(\psi(a))uu^{-1}\xi(v)u^{-1} \\
&= \psi(a)\omega q(u)^{-1}\theta(\psi(a), u) \cdot u^{-1}\xi(v)u^{-1}
\end{aligned}$$

and therefore

$$\begin{aligned}
\psi(a)M(a, v)\xi(v)u^{-1} &= \psi(a)M(a, 1)u \cdot u^{-1}\xi(v)u^{-1} \\
&= \psi(a)M(a, 1)\xi(v)u^{-1}
\end{aligned}$$

for all $a \in X$ and all $v \in L$. It follows that $M(a, v)$ is independent of v. Let $M(a) = M(a, 1)$ for all $a \in X$. Thus

$$(12.13) \qquad \xi(\theta(a, v)) = \omega q(u)^{-1} \big(\theta(\psi(a), \xi(v)) + M(a)\xi(v) \big)$$

for all $a \in X$ and all $v \in L$ (by 12.12).

Next we observe that

$$f(\xi(\pi(a)), \xi(\delta)) = q(u)f(\pi(a), \delta) = 0$$

and

$$(12.14) \qquad f(\xi(1), \xi(\delta)) = q(u)$$

for all $a \in X$ by 4.1 and 12.5.i. By 12.13 with $v = 1$, therefore,

$$f(\theta(\psi(a), u), \xi(\delta)) + q(u)M(a) = 0$$

and hence

$$(12.15) \qquad M(a) = -q(u)^{-1} f(\theta(\psi(a), u), \xi(\delta))$$

for all $a \in X$. Note that by 4.1, $\xi(\delta) = u/2$ and

$$f(\pi(\psi(a)), 1) = 0,$$

and thus (by 4.9.i and 12.15) M is, in fact, identically zero if $\mathrm{char}(K) \neq 2$. In any event, we have

$$\begin{aligned}
M(a + b) &- M(a) - M(b) \\
&= -q(u)^{-1} f\big(h(\psi(a), \psi(b)u) + g(\psi(a), \psi(b))u, \xi(\delta)\big) \\
&\qquad\qquad\qquad\qquad\qquad\qquad \text{by C3 and 12.15} \\
&= -\omega^{-1} f\big(\xi(h(a, b)), \xi(\delta)\big) + g(\psi(a), \psi(b)) \\
&\qquad\qquad\qquad\qquad\qquad\qquad \text{by 12.6 and 12.14} \\
&= -q(u)\omega^{-1} f(h(a, b), \delta) + g(\psi(a), \psi(b)) &\text{by 12.5.i} \\
&= -q(u)\omega^{-1} g(a, b) + g(\psi(a), \psi(b)) &\text{by 4.3}
\end{aligned}$$

and

$$\begin{aligned}
\phi(\psi(a + b), u) &= \phi(\psi(a), u) + \phi(\psi(b), u) \\
&\quad + g(\psi(a)u, \psi(b)u) - g(\psi(a), \psi(b))q(u) \quad \text{by 4.18}
\end{aligned}$$

for all $a, b \in X$, from which it follows that the maps

$$(a, t) \mapsto (\psi(a), tq(u)\omega^{-1} + M(a))$$

and

$$(a, t) \mapsto (\psi(a)u, tq(u)^2\omega^{-1} + M(a)q(u) + \phi(\psi(a), u))$$

are both automorphisms of the group S defined in 11.10. By 12.15, we also have

$$(12.16) \qquad M(0) = 0.$$

Recall, too, that $\phi(0, L) = 0$ by 3.12. Suppose as well that

$$(12.17) \qquad \begin{aligned} M(av) + M(a)q(v) &= q(u)^{-1}\big(\phi(\psi(a), \xi(v)) + \phi(\psi(av), u)\big) \\ &\quad + \phi(a, v)q(u)\omega^{-1} \end{aligned}$$

for all $a \in X$ and all $v \in L$. By 12.5, 12.13, 12.16 and 12.17, the maps

$$\begin{aligned}
x_1(a, t) &\mapsto x_1(\psi(a), tq(u)\omega^{-1} + M(a)), \\
x_2(v) &\mapsto x_2(\xi(v)q(u)\omega^{-1}), \\
x_3(a, t) &\mapsto x_3(\psi(a)u, tq(u)^2\omega^{-1} + M(a)q(u) + \phi(\psi(a), u)), \\
x_4(v) &\mapsto x_4(\xi(v))
\end{aligned}$$

preserve the relations (in 11.11.ii) defining Ω_Ξ. By 11.4, therefore, there exists an element ν of H_0 whose image in $\mathrm{Str}(\Xi)$ is (ξ, ψ) (under the map defined at the beginning of this proof).

It thus remains only to verify the identity 12.17 (but this turns out to be surprisingly difficult). The maps M and ϕ are identically zero if $\operatorname{char}(K) \neq 2$. We can assume, therefore, that $\operatorname{char}(K) = 2$. Choose $a \in X$ and $v \in L$. Then

$$
\begin{aligned}
M(av) &= q(u)^{-1} f\big(\theta(\psi(av), u), \xi(\delta)\big) && \text{by 12.15} \\
&= q(u)^{-1} f\big(\theta(\psi(a)\xi(v)u^{-1}, u), \xi(\delta)\big) && \text{by 12.5.ii} \\
&= q(u)^{-3} f\big(\theta(\psi(a)\xi(v)u^\sigma, u), \xi(\delta)\big) && \text{by C2.}
\end{aligned}
$$
(12.18)

By 4.24, we have

$$
\begin{aligned}
\theta(\psi(a)\xi(v)u^\sigma, u) &= \theta(\psi(a)\xi(v), u^\sigma)^\sigma q(u) \\
&\quad + f\big(\theta(\psi(a)\xi(v), u^\sigma), u^\sigma\big)u + \phi(\psi(a)\xi(v), u^\sigma)u.
\end{aligned}
$$
(12.19)

Next we observe that

$$
\begin{aligned}
f(\theta(\psi(a)\xi(v), u^\sigma), u^\sigma) &= f(\pi(\psi(a)\xi(v)), 1)q(u) && \text{by 3.19.i} \\
&= f(\pi(\psi(a)), 1)q(\xi(v))q(u) && \text{by 3.21} \\
&= f(\pi(\psi(a)), 1)q(u)^2 q(v) && \text{by 12.5.i}
\end{aligned}
$$

and (since $q(\xi^{-1}(v)) = q(u)^{-1}q(v)$ for all $v \in L$ by 12.5.i and $f(1,1) = 0$)

$$
\begin{aligned}
f(\pi(\psi(a)), 1) &= f\Big(\theta\big(\psi(a), \xi(\xi^{-1}(1))\big), 1\Big) \\
&= q(u)\omega^{-1} f\Big(\xi\big(\theta(a, \xi^{-1}(1))\big), \xi(\xi^{-1}(1))\Big) && \text{by 12.13} \\
&= q(u)^2 \omega^{-1} f\big(\theta(a, \xi^{-1}(1)), \xi^{-1}(1)\big) && \text{by 12.5.i} \\
&= q(u)\omega^{-1} f(\pi(a), 1) && \text{by 3.19.i,}
\end{aligned}
$$

so

$$
f\big(\theta(\psi(a)\xi(v), u^\sigma), u^\sigma\big) = q(u)^3 \omega^{-1} q(v) f(\pi(a), 1).
$$
(12.20)

We also have

$$
\begin{aligned}
\phi(\psi(a)\xi(v), u^\sigma) &= \phi(\psi(av)u, u^\sigma) && \text{by A3 and 12.5.ii} \\
&= \phi(\psi(av)u, u^{-1})q(u)^2 && \text{by 4.14} \\
&= \phi(\psi(av), u)q(u) && \text{by 4.23.}
\end{aligned}
$$
(12.21)

By C4 and 12.5.i, we have

$$
\begin{aligned}
\theta(\psi(a)\xi(v), u^\sigma)^\sigma &= \theta(\psi(a), u)q(\xi(v)) + f(u, \xi(v))\theta(\psi(a), \xi(v)) \\
&\quad + f(\theta(\psi(a), \xi(v)), u)\xi(v) + \phi(\psi(a), \xi(v))u \\
&= \theta(\psi(a), u)q(u)q(v) + f(1, v)q(u)\theta(\psi(a), \xi(v)) \\
&\quad + f(\theta(\psi(a), \xi(v)), u)\xi(v) + \phi(\psi(a), \xi(v))u
\end{aligned}
$$

and therefore

$$
\begin{aligned}
&f\big(\theta(\psi(a)\xi(v), u^\sigma)^\sigma, \xi(\delta)\big) \\
&= q(u)\Big(f(\theta(\psi(a), u), \xi(\delta))q(v) + f(v, 1)f(\theta(\psi(a), \xi(v)), \xi(\delta)) \\
&\quad + f(\theta(\psi(a), \xi(v)), u)f(v, \delta) + \phi(\psi(a), \xi(v))\Big).
\end{aligned}
$$
(12.22)

Substituting

$$f(\theta(\psi(a), u), \xi(\delta)) = q(u)M(a) \qquad \text{by 12.15}$$

as well as

$$\begin{aligned}
f(\theta(\psi(a), \xi(v)), \xi(\delta)) &= f(\xi(v), \xi(\delta))M(a) \\
&\quad + q(u)\omega^{-1}f(\xi(\theta(a, v)), \xi(\delta)) \\
&\qquad\qquad\qquad\qquad \text{by 12.13} \\
&= q(u)f(v, \delta)M(a) + q(u)^2\omega^{-1}f(\theta(a, v), \delta)
\end{aligned}$$

and

$$\begin{aligned}
f(\theta(\psi(a), \xi(v)), u) &= f(\xi(v), u)M(a) \\
&\quad + q(u)\omega^{-1}f(\xi(\theta(a, v)), u) \\
&\qquad\qquad\qquad\qquad \text{by 12.13} \\
&= q(u)f(v, 1)M(a) + q(u)^2\omega^{-1}f(\theta(a, v), 1)
\end{aligned}$$

in 12.22, we obtain

$$\begin{aligned}
&f\big(\theta(\psi(a)\xi(v), u^\sigma)^\sigma, \xi(\delta)\big) \\
&= M(a)q(u)^2q(v) + q(u)^3\omega^{-1}\Big(f(\theta(a, v), \delta)f(v, 1) + f(\theta(a, v), 1)f(v, \delta)\Big) \\
&\quad + \phi(\psi(a), \xi(v))q(u).
\end{aligned}$$

With v^* as in 4.11, we have

$$\begin{aligned}
f(\theta(a, v), \delta)f(v, 1) + f(\theta(a, v), 1)f(v, \delta) & \\
= f(\theta(a, v), v + v^*) &\qquad \text{by C1} \\
= f(\theta(a, v), v) + \phi(a, v) &\qquad \text{by 4.12} \\
= f(\pi(a), 1)q(v) + \phi(a, v) &\qquad \text{by 3.19.i.}
\end{aligned}$$

Therefore

$$\begin{aligned}
&f\big(\theta(\psi(a)\xi(v), u^\sigma)^\sigma, \xi(\delta)\big)q(u) \\
\text{(12.23)} \qquad &= M(a)q(u)^3q(v) + q(u)^4\omega^{-1}\big(f(\pi(a), 1)q(v) + \phi(a, v)\big) \\
&\quad + q(u)^2\phi(\psi(a), \xi(v))
\end{aligned}$$

By 12.18, 12.19, 12.20, 12.21 and 12.23, we conclude that

$$\begin{aligned}
q(u)^3M(av) &= M(a)q(u)^3q(v) + q(u)^4\omega^{-1}\phi(a, v) \\
&\quad + q(u)^2\phi(\psi(a), \xi(v)) + \phi(\psi(av), u)q(u)^2.
\end{aligned}$$

Thus 12.17 holds. $\qquad\qquad\qquad\qquad\qquad\qquad\qquad\qquad\qquad\quad \square$

Bibliography

[1] De Medts, T. 2002 A characterization of quadratic forms of type E_6, E_7 and E_8, *J. Algebra* **22**, 394-410.

[2] De Medts, T. 2004 Automorphisms of F_4 quadrangles, *Math. Annalen* **328**, 399-413.

[3] De Medts, T. 2004 Quadrangular systems, *Memoirs Amer. Math. Soc.*, vol. 173, nr. 818.

[4] Herstein, I. 1976 *Rings with Involution*, University of Chicago Press, Chicago, London.

[5] Jacobson, N. 1956 *The Structure of Rings*, Amer. Math. Soc., Providence.

[6] Jacobson, N. and K. McCrimmon 1971 Quadratic Jordan algebras of quadratic forms with basepoint, *J. Indian Math. Soc.* **35**, 1-45.

[7] Jordan, P., von Neumann, J. and Wigner, E. 1934 On an algebraic generalization of the quantum mechanical formalism, *Ann. Math.* **35**, 29-64.

[8] Petersson, H. P. and Racine, M. L. 1986 Jordan algebras of degree 3 and the Tits process, *J. Algebra* **98**, 211-243.

[9] Petersson, H. P. and Racine, M. L. 1986 Classification of algebras arising from the Tits process, *J. Algebra* **98**, 244-279.

[10] Tits, J. 1966 Classification of algebraic simple groups, in *Algebraic Groups and Discontinuous Groups, Boulder, 1965, Proc. Symp. Pure Math.* **9**, Amer. Math. Soc., Providence, pp. 33-62.

[11] Tits, J. 1974 *Buildings of Spherical Type and Finite BN-Pairs*, Lecture Notes in Mathematics, vol. 386, Springer, Berlin, Heidelberg, New York.

[12] Tits, J. and Weiss, R. M. 2002 *Moufang Polygons*, Springer, Berlin, Heidelberg, New York.

[13] Weiss, R. M. 2004 *The Structure of Spherical Buildings*, Princeton University Press, Princeton.

[14] Weiss, R. M. 2005 Moufang quadrangles of type E_6 and E_7, *J. reine u. angew. Math.*, to appear.

Index

alternative division ring, vii
anisotropic
  kernel, ix, 119
  pseudo-quadratic space, 3
  quadratic space, 1
apartment, 109

basepoint, 1, 17

Clifford algebra
  even, 18
  with basepoint, 17

De Medts, T., ix
defect, 11
defective
  quadrangular algebra, 9
  quadratic space, 11
degenerate, 11
δ-standard, 29
Dieudonné, J., 3

e-orthogonal, 46
equivalent, 7
exceptional
  Moufang quadrangle, viii
  quadrangular algebra, 9

generalized polygon, 109

hexagonal system, vii

improper, 83
indifferent set, 91
involution
  of a ring, 2
  standard, 1, 2
isomorphic
  pointed quadratic spaces, 9
  quadrangular algebras, 9
  quadratic spaces, 11
  root group sequences, 111
isotope, 81

Jacobson, N., 17

maximal unipotent subgroup, ix, 110

McCrimmon, K., 17
Moufang condition, 109

norm splitting map, 15

orthogonal sum, 12

pointed quadratic space, 1
polygonal sequence, 109
proper, 9
pseudo-quadratic space, 3
  anisotropic, 3
  standard, 3

quadrangular algebra, 4
quadratic
  Jordan algebra, vii
  skew-field with involution, 3
quadratic space, 1
  anisotropic, 1
  of type E_6, E_7 and E_8, 14
  of type F_4, 15
  pointed, 1
quaternion algebra, 2

radical, 11
reduced norm, 2
regular, 9
root, 109
root group, 109
root group sequence, 110
  opposite, 111

separable plane, 14
similar quadratic spaces, 11
skew-hermitian form, 3
special, 9
standard
  pseudo-quadratic space, 3
  quadrangular algebra, 29
standard involution
  of a quadratic space, 1
  of a quaternion algebra, 2
structure group, 127

Tits, J., vii
T-orthogonal, 16
traces, 2